TECHNICAL MANUAL

# ARMY AMMUNITION DATA SHEETS SMALL CALIBER AMMUNITION: Technical Manual TM 43-0001-27 C2

PrepperPress
*Post-Apocalyptic Fiction & Survival Nonfiction*

DEPARTMENT OF THE ARMY - APRIL 1994

Published by Prepper Press
ISBN # 978-0615510170

Technical Manual )
)
No. 43-0001-27 )

HEADQUARTERS
DEPARTMENT OF THE ARMY
Washington, DC, 29 April 1994

**ARMY AMMUNITION DATA SHEETS**
**SMALL CALIBER AMMUNITION**
**FSC 1305**

REPORTING OF ERRORS AND RECOMMENDING IMPROVEMENTS

You can improve this manual. If you find any mistakes or know of a way to improve the procedures, please let us know. Mail your DA Form 2028 (Recommended Changes to Publications and Blank Forms), or DA Form 2028-2 located in the back of this manual direct to Commander, U.S. Army Armament Research, Development and Engineering Center; ATTN: SMCAR-LSB, Picatinny, Arsenal, NJ 07806-5000. A reply will be furnished to you.



This manual supersedes TM 43-0001-27, 29 June 1981, including all changes.

THIS PAGE INTENTIONALLY LEFT BLANK

# CHAPTER 1
# INTRODUCTION

## 1-1. PURPOSE

This manual is a reference handbook published as an aid in planning, training, familiarization and identification of small arms ammunition, ranging from 22 caliber through 30 millimeter and shotgun ammunition from .410 to 10 gage. This is the Federal Supply Class 1305. Operating instructions are contained in the appropriate weapons manuals.

## 1-2. SCOPE

a. For each item of materiel, there are illustrations and descriptions together with characteristics and related data. Included in the related data are weights, dimensions, performance data, shipping and storage data, type classification, and logistics control code (LCC). Refer to SC 1305/30-IL for weight, dimensions and cube of packing boxes for small arms ammunition.

b. Information concerning supply, operation, and maintenance of the items will be found in the publications referenced for those items. A complete listing of these publications is maintained in DA Pam 310 series indexes.

c. Within this manual, items with the following type classifications are included:

(1) Standard (STD)

(2) Contingency (CON)

(3) Limited Procurement (LP)

(4) Reclassified obsolete (OBS) for regular Army use, but used by National Guard or Reserve Units.

(5) Reclassified . OBS for all Army use, but used by Marine Corps, Air Force or Navy.

(6) Reclassified OBS, no users, but U.S. stocks remain.

d. Items with the following type classification are not included: Reclassified OBS for all U.S. use. No U.S. stocks remain. (Foreign use or stock may remain.)

e. Numerical values, such as weights, dimensions, candlepower, etc., are nominal values, except when specified as maximum or minimum. Actual items may vary slightly from these values. Allowable limits can be obtained from the drawings indicated in the data sheets.

## 1-3. KEY TO ABBREVIATIONS AND SYMBOLS

A .................................. Army
AF .................................. Air Force
band .................................. bandoleer
belt .................................. linked ammo (cloth or metal)
bx .................................. box
cal .................................. caliber
cm .................................. centimeter(s)
comm .................................. commercial
ctg .................................. cartridge
ctn .................................. carton
DODIC .................................. Department of Defense Identification Code
DODAC .................................. Department of Defense Ammunition Code
ft .................................. foot/feet
fps .................................. feet per second
g .................................. gram(s)
gr .................................. grain(s)
GSA .................................. General Services Administration
H .................................. Height
in .................................. inch(es)
L .................................. Length
m .................................. meter(s)
MBR .................................. Meet Ballistic Requirement
MC .................................. Marine Corps
mm .................................. millimeter(s)
MPa .................................. Mega Pascal
mps .................................. meters per second
mtl bx .................................. metal box
mtl lnd bx .................................. metal lined box
N .................................. Navy
NATO .................................. North Atlantic Treaty Organization
SCG .................................. Storage Compatibility Group
psi .................................. pounds per square inch
W .................................. Width
wdn bx .................................. wooden box
wrbnd bx .................................. wirebound box
yd .................................. yard(s)

## 1-4. METRIC CONVERSION CHART

For approximate conversions to/from metric measures see table 1-1.

### Table 1-1. Metric Conversion Chart

Approximate Conversions to Metric Measures

| Symbol | When You Know | Multiply By | To Find | Symbol |
|---|---|---|---|---|
| | | **LENGTH** | | |
| in. | inches | 2.5 | centimeters | cm |
| ft | feet | 30 | centimeters | cm |
| yd | yards | 0.9 | meters | m |
| mi | miles | 1. | kilometers | km |
| | | **AREA** | | |
| $in^2$ | square inches | 6.5 | square centimeters | $cm^2$ |
| ft2 | square feet | 0.09 | square meters | m2 |
| $yd^2$ | square yards | 0.8 | square meters | $m^2$ |
| mil | square miles | 2.6 | square kilometers | $km^2$ |
| | acres | 0.4 | hectares | ha |
| | | **WEIGHT** | | |
| oz | ounces | 28 | grams | g |
| lb | pounds | 0.45 | kilograms | kg |
| | short tons (2000 lb) | 0.9 | tonnes | t |
| | | **VOLUME** | | |
| tsp | teaspoons | 5 | milliliters | ml |
| Tbsp | tablespoons | 15 | milliliters | ml |
| fl oz | fluid ounces | 30 | milliliters | ml |
| c | cups | 0.24 | liters | 1 |
| pt | pints | 0.47 | liters | 1 |
| qt | quarts | 0.95 | liters | 1 |
| gal | gallons | 3.8 | liters | 1 |
| $ft^3$ | cubic feet | 0.03 | cubic meters | $m^2$ |
| $yd^3$ | cubic yards | 0.70 | cubic meters | $m^3$ |

**TEMPERATURE**

| Symbol | When You Know | Subtract | Multiply By | To Find | Symbol |
|---|---|---|---|---|---|
| °F | Fahrenheit | 32 | 0.55 | Celsius | °C |

Approximate Conversions from Metric Measures

| Symbol | When You Know | Multiply By | To Find | Symbol |
|---|---|---|---|---|
| | | **LENGTH** | | |
| mm | millimeters | 0.04 | inches | in. |
| cm | centimeters | 0.4 | inches | in. |
| m | meters | 3.3 | feet | ft |
| m | meters | 1.1 | yards | yd |
| km | kilometers | 0.6 | miles | mi |
| | | **AREA** | | |
| $cm^2$ | square centi-meters | 0.16 | square inches | $in^2$ |
| $m^2$ | square meters | 1.2 | square yards | $yd^2$ |
| $km^2$ | square kilo-meters | 0.4 | square miles | $mi^2$ |
| ha | hectares (10,000 $m^2$) | 2.5 | acres | |
| | | **WEIGHT** | | |
| g | grams | 0.035 | ounces | oz |
| kg | kilograms | 2.2 | pounds | lb |
| t | tonnes (1000 kg) | 1.1 | short tons | |
| | | **VOLUME** | | |
| ml | milliliters | 0.03 | fluid ounces | fl oz |
| l | liters | 2.1 | pints | pt |
| l | liters | 1.06 | quarts | qt |
| l | liters | 0.26 | gallons | gal |
| $m^3$ | cubic meters | 35 | cubic feet | $ft^3$ |
| $m^3$ | cubic meters | 1.3 | cubic yards | $yd^3$ |

**TEMPERATURE**

| Symbol | When You Know | Multiply By | Add | To Find | Symbol |
|---|---|---|---|---|---|
| °C | Celsius | 1.8 | 32 | Fahrenheit | °F |

# CHAPTER 2

# SHOTGUN CARTRIDGES

THIS PAGE INTENTIONALLY LEFT BLANK

## CARTRIDGE, .410, SHOTGUN, NO. M35

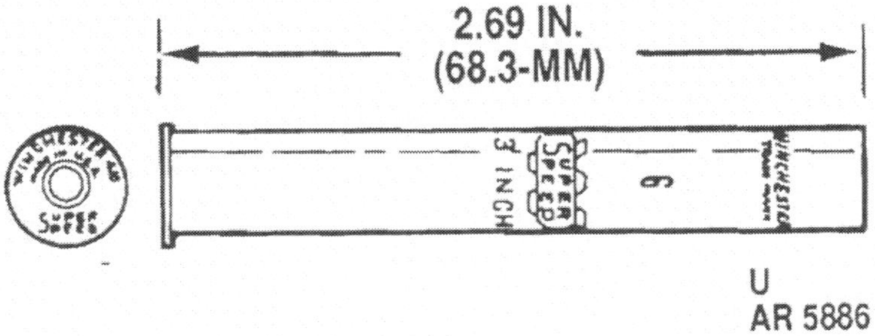


### Type Classification:

OBS - MSR 11756003.

### Use:

Rifle/Shotgun, Caliber .22/.410 Bore, Survival, M6. The cartridge is intended for use in survival weapons against small game.

### Description:

The cartridge case is all aluminum, and is loaded with smokeless powder and No. 6 copper-coated lead shot.

### Tabulated Data:

DODAC ................ 1305-A055
UNO serial number ............. 0012
UNO proper shipping
  name ................ Cartridges for weapons, inert projectile
Weight ................ 430 gr
Length ................ 2.69 in. (68.3 mm)
Tracer ................ NA
Primer ................ Percussion
Fuze ................ NA

Explosive:
  Type ................ NA
  Weight ................ NA
Incendiary:
  Type ................ NA
  Weight ................ NA
Propellant:
  Type ................ Smokeless powder
  Weight ................ 7 gr

### Performance:

Chamber pressure ............... 13,000 psi
Velocity ................ 960 fps, 3 ft from muzzle

Shipping and Storage Data:
Quantity-distance class/
  SCG ................ 1.4S
Storage code ................ Class V
DOT shipping class ............. C
DOT designation ................ SMALL ARMS AMMUNITION
Drawing number ................ 7553403

### References:

SB 700-20
TM 9-1300-206

THIS PAGE INTENTIONALLY LEFT BLANK

## CARTRIDGE, 10 GAGE, SHOTGUN, BLANK

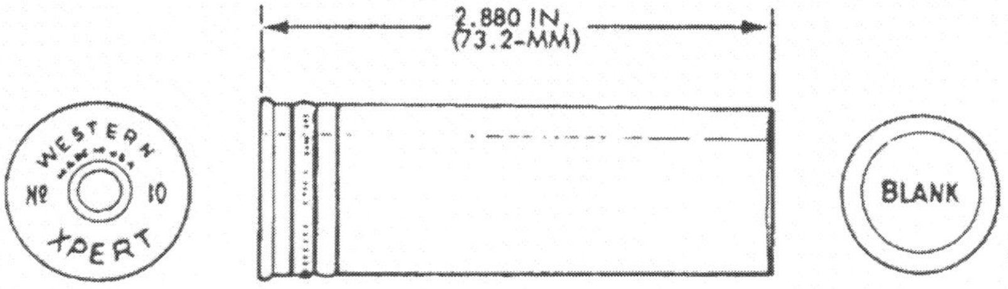


U
AR 5887

**Type Classification:**

STD - OTCM 36524.

**Use:**

This is an optional use item for reveille and retreat salutes, and is fired in the 3-inch gun, 75mm gun, 75mm howitzer, or 105lmm howitzer.

**Description:**

The cartridge is similar to standard shotgun cartridges but it contains no lead shot. It has a paper cartridge case. The forward closing disc is marked "BLANK."

**Function:**

The cartridge is designed to produce a noise when initiated. It is used as a salute item in large caliber weapons. The blank is inserted either in a prepared cartridge case or in a breech block of the weapon being used.

**Tabulated Data:**

DODAC................................1305-A010
UNO serial number .............0014
UNO proper shipping
 name ..............................Cartridges for weapons, blank
Weight ................................290 gr
Length.................................2.88 in. (73.2 mm)
Tracer .................................NA
Primer.................................Percussion
Fuze...................................NA
Explosive:
 Type ...............................NA
 Weight.............................NA
Incendiary:
 Type ...............................NA
 Weight.............................NA
Propellant:
 Type ............................... Black powder
 Weight.............................8 g

**Performance:**

Chamber pressure .............NA
Velocity...............................NA

**Shipping and Storage Data:**

Quantity-distance class/
 SCG.................................1.4S
Storage code ......................Class V
DOT shipping class ............C
DOT designation ................SMALL ARMS AMMUNITION
Drawing number .................10524632

**References:**

SB 700-20
TM 9-1300-206

THIS PAGE INTENTIONALLY LFET BLANK

## CARTRIDGE, 12 GAGE, SHOTGUN, NO. 00

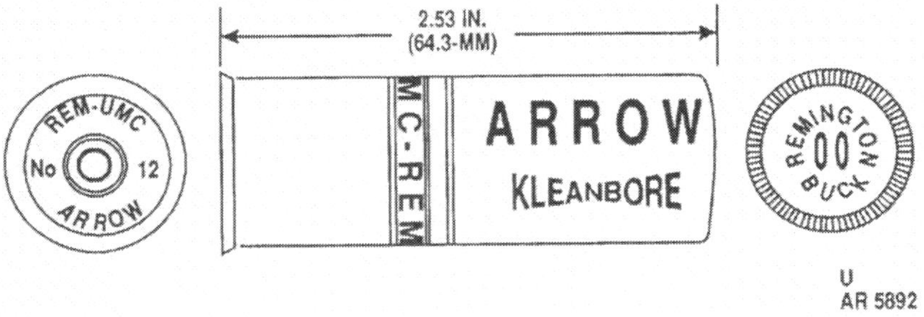


**Type Classification:**

STD - OTCM 36841.

**Use:**

Military issue shotgun, 2-3/4 inch chamber. The cartridge is intended for guard and combat use.

**Description:**

The cartridge case is paper, and is loaded with smokeless powder and No. 00 commercial shot.

**Tabulated Data:**

DODAC ............................... 1305-A011
UNO serial number ............. 0012
UNO proper shipping
name ............................... Cartridges for weapons, inert projectile
Weight ................................ 791 gr
Length ................................ 2.53 in. (64.3 mm)
Tracer ................................ NA
Primer ................................ Percussion
Fuze ................................... NA

Explosive:
Type ................................ NA
Weight ................................ NA
Incendiary:
Type ................................ NA
Weight ............................. NA
Propellant:
Type ................................ Smokeless powder
Weight ............................. MBR

**Performance:**

Chamber pressure ............... 11,000 psi
Velocity .............................. 1325 fps, 3 ft from muzzle

**Shipping and Storage Data:**

Quantity-distance class/
SCG ................................ 1.4S
Storage code ....................... Class V
DOT shipping class ............. C
DOT designation ................. SMALL ARMS AMMUNITION
Drawing number .................. 7553929

**References:**

SB 700-20
TM 9-1005-303-14
TM 9-1300-206

THIS PAGE INTENTIONALLY LEFT BLANK

## CARTRIDGE, 12 GAGE, SHOTGUN, NO. 00, M162

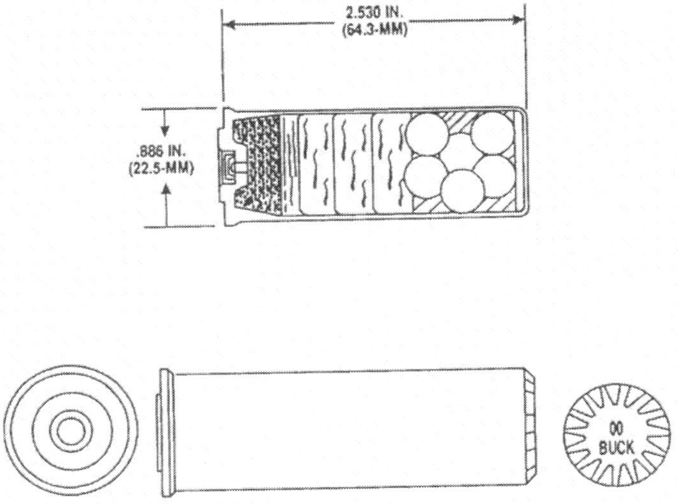


U
AR 5891

**Type Classification:**

LP - OTCM 38036.

**Use:**

Military issue shotgun, 2-3/4 inch chamber. The cartridge is intended for guard and combat use.

**Description:**

The cartridge case is all plastic, and is loaded with smokeless powder and No. 00 commercial shot.

**Tabulated Data:**

DODAC ..............................1305-A011
UNO serial number ............0012
UNO proper shipping
  name ..............................Cartridges for weapons, inert projectile
Weight ...............................0.736 gr
Length ................................2.53 in. (64.3 mm)
Tracer .................................NA
Primer ................................Percussion
Fuze....................................NA
Explosive:
  Type ................................NA
  Weight .............................NA
Incendiary:
  Type ................................NA
  Weight .............................NA
Propellant:
  Type ...............................Smokeless powder
  Weight ............................MBR

**Performance:**

Chamber pressure ..............11,000 psi
Velocity ..............................1325 fps, 3 ft from muzzle

**Shipping and Storage Data:**

Quantity-distance class/
  SCG.................................1.4S
Storage code.......................Class V

DOT shipping class ............C
DOT designation ................SMALL ARMS AMMUNITION
Drawing number .................10522423

**References:**

SB 700-20
TM 9-1005-303-14
TM 9-1300-206

## CARTRIDGE, 12 GAGE, SHIOTGUN, NO. M19

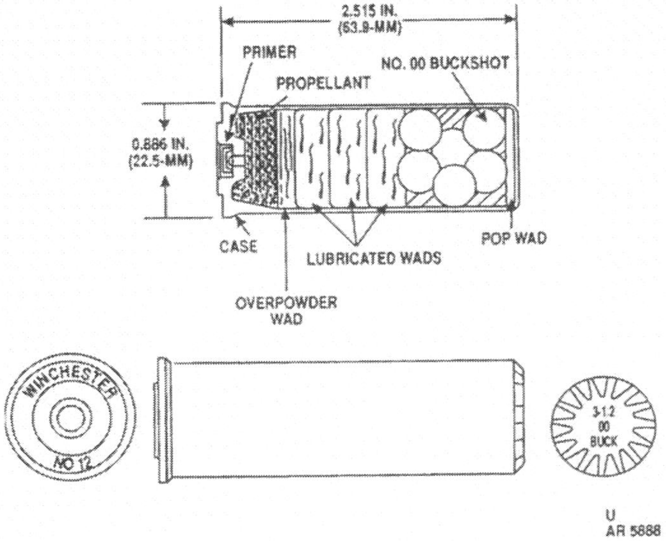


**Type Classification:**

STD - OTCM 36841.

**Use:**

Military issue shotgun, 2-3/4 inch chamber. The cartridge is intended for guard and combat use.

**Description:**

The cartridge case is all brass, and is loaded with smokeless powder and No. 00 commercial shot.

**Tabulated Data:**

DODAC ................................1305-A011
UNO serial number ............0012
UNO proper shipping
name ..............................Cartridges for weapons, inert projectile
Weight ...............................930 gr
Length ................................2.515 in. (63.9 mm)
Tracer ................................NA
Primer ................................Percussion
Fuze ...................................NA
Explosive:
Type ...............................NA
Weight ............................NA
Incendiary:
Type ...............................NA
Weight ............................NA
Propellant:
Type ...............................Smokeless powder
Weight ............................26 gr

**Performance:**

Chamber pressure ..............11,000 psi
Velocity .............................1125 fps, 3 ft from muzzle

**Shipping and Storage Data:**

Quantity-distance class/
SCG................................1.4S
Storage code ...................... Class V

DOT shipping class ............C
DOT designation ................SMALL ARMS AMMUNITION
Drawing number .................7640981

**References:**

SB 700-20
TM 9-1005-303-14
TM 9-1300-206

## CARTRIDGE, 12 GAGE, SHOTGUN, NO. M257

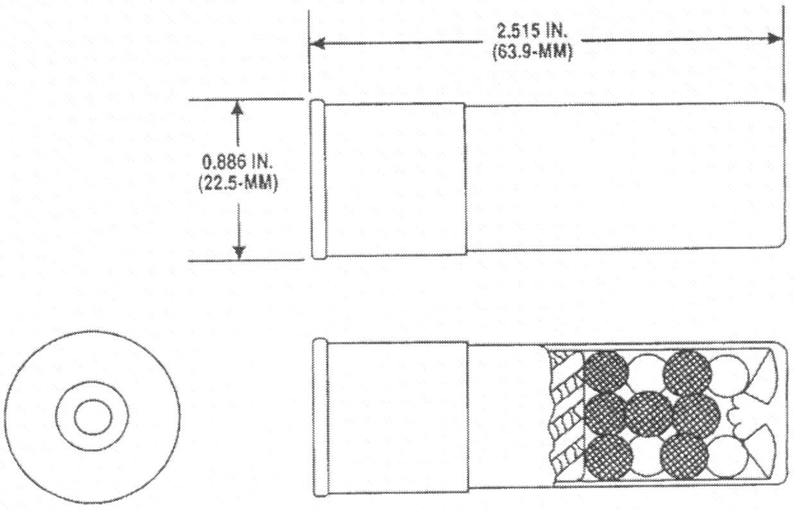


**Type Classification:**

Terminated - MSR 08752013.

**Use:**

Military issue shotgun, 20-inch full choke barrel. The cartridge is intended for guard duty and for riot control weapons.

**Description:**

The cartridge case is plastic, and is loaded with smokeless powder and No. 4 hard lead antimony pellets.

**Tabulated Data:**

| | |
|---|---|
| DODAC | 1305-A011 |
| UNO serial number | 0012 |
| UNO proper shipping name | Cartridges for weapons, inert projectile |
| Weight | 748 gr |
| Length | 2.515 in. (63.9 mm) |
| Tracer | NA |
| Primer | Percussion |
| Fuze | NA |
| Explosive: | |
| Type | NA |
| Weight | NA |
| Incendiary: | |
| Type | NA |
| Weight | NA |
| Propellant: | |
| Type | Smokeless powder |
| Weight | MBR |

**Performance:**

| | |
|---|---|
| Chamber pressure | 11,000 psi |
| Velocity | 1335 fps, 3 ft from muzzle |

**Shipping and Storage Data:**

| | |
|---|---|
| Quantity-distance class/ SCG | 1.4S |
| Storage code | Class V |

DOT shipping class ............C
DOT designation ................SMALL ARMS AMMUNITION
Drawing number .................10542446

**References:**

SB 700-20
TM 9-1005-303-14
TM 9-1300-206

## CARTRIDGE, 12 GAGE, SHOTGUN, NO. M274

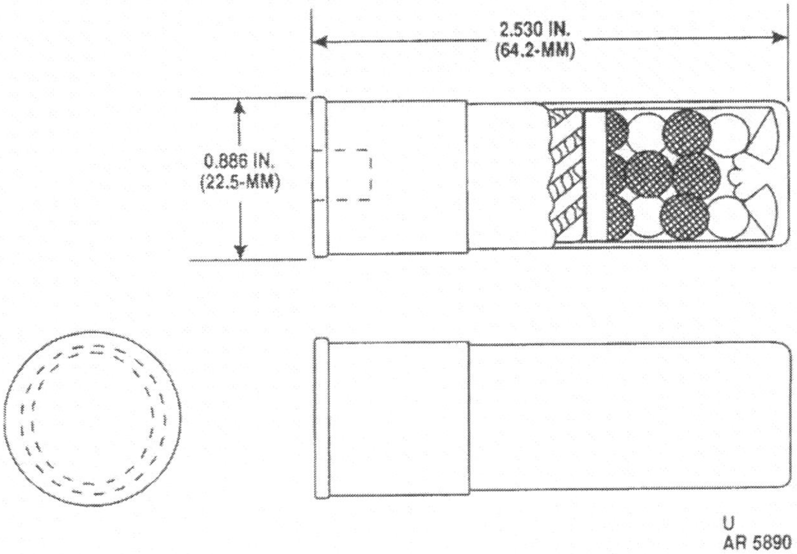


**Type Classification:**

OBS - MSR 11756003.

**Use:**

Military issue, riot-type shotgun, 20-inch barrel cylinder bore. The cartridge is intended for use against small game and for riot control weapons.

**Description:**

The cartridge case may be paper or plastic, and is loaded with smokeless powder and No. 4 hard chilled shot.

**Tabulated Data:**

DODAC ............ 1305-A011
UNO serial number ............ 0012
UNO proper shipping name ............ Cartridges for weapons, inert projectile
Weight ............ 740 gr
Length ............ 2.53 in. (64.2 mm)
Tracer ............ NA
Primer ............ Percussion
Fuze ............ NA
Explosive:
  Type ............ NA
  Weight ............ NA
Incendiary:
  Type ............ NA
  Weight ............ NA
Propellant:
  Type ............ Smokeless powder
  Weight ............ MBR

**Performance:**

Chamber pressure ............ 11,000 psi
Velocity ............ 1255 fps, 3 ft from muzzle

**Shipping and Storage Data:**

Quantity-distance class/SCG ............ 1.4S
Storage code ............ Class V

DOT shipping class ............ C
DOT designation ................ SMALL ARMS AMMUNITION
Drawing number ................. 10542495

**References:**

SB 700-20
TM 9-1005-303-14
TM 9-1300-206

# CHAPTER 3

# CALIBER .22 CARTRIDGES

THIS PAGE INTENTIONALLY LEFT BLANK

## CARTRIDGE, CALIBER .22, BALL, LONG RIFLE

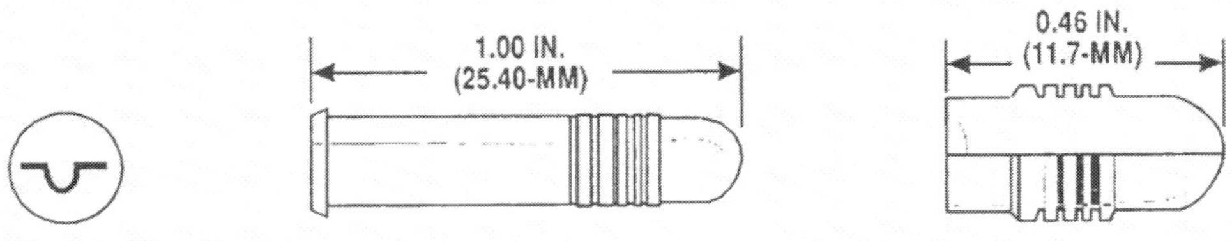


U
AR 5895

**Type Classification:**

STD - OTCM 36841.

**Use:**

Subcaliber Rifle M2A1; Winchester Models 52 and 75; Remington Models M40X and 513T; Machine Gun Trainers M3 and M4. The cartridge is intended for marksmanship practice and match use.

**Description:**

BALL Cartridge. The cartridge is grade match.

**Tabulated Data:**

DODAC .................................... 1305-A086
UNO serial number .................. 0012
UNO proper shipping
name .... .............................. Cartridges for weapons, inert projectile
Weight .................................... 52 gr
Length...................................... 1 in. (25.4 mm)
Tracer ...................................... NA
Primer ..................................... Percussion (Rimfire)
Fuze NA
Explosive:
Type .................................... NA
Weight ................................. NA
Incendiary:
Type .................................... NA
Weight ................................. NA
Propellant:
Type .................................... Smokeless powder
Weight ................................. MBR

**Performance:**

Chamber pressure .................... 24,000 psi
Velocity .................................... 1100 fps, 15 ft from muzzle

**Shipping and Storage D ata:**

Quantity-distance class/
SCG ..................................... 1.4S
Storage code ............................ Class V
DOT shipping class .................. C
DOT designation ...................... SMALL ARMS AMMUNITION
Drawing number ....................... 7553931

**References:**

SB 700-20
TM 9-1005-206-14P/4
TM 9-1005-226-14
TM 9-1300-206

THIS PAGE INTENTIONALLY LEFT BLANK

CARTRIDGE, CALIBER .22, BALL, LONG RIFLE, M24

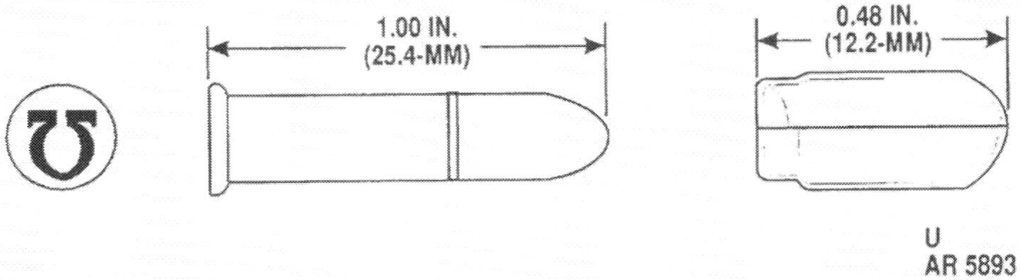


**Type Classification:**

OBS - MSR 11756003.

**Use:**

Subcaliber Rifle M2A1; Caliber .22 Rifle; Remington Models 40X and M513T; Steven's Model 416-2; Winchester Models 52 and 75; Machine Gun Trainers M3 and M4; pistols for gallery practice and training purposes. The cartridge is intended for use against small game for survival purposes.

**Description:**

BALL Cartridge.

**Tabulated Data:**

| | |
|---|---|
| DODAC | 1305-A086 |
| UNO serial number | 0012 |
| UNO proper shipping name | Cartridges for weapons, inert projectile |
| Weight | 53 gr |
| Length | 1 in. (25.4 mm) |
| Tracer | NA |
| Primer | Percussion (Rimfire) |
| Fuze | NA |
| Explosive: | |
| Type | NA |
| Weight | NA |
| Incendiary: | |
| Type | NA |
| Weight | NA |
| Propellant: | |
| Type | Smokeless powder |
| Weight | 2.5 gr |
| Projectile: | |
| Weight | 40.5 gr |

**Performance:**

| | |
|---|---|
| Chamber pressure | 24,000 psi |
| Velocity | 1185 fps, 15 ft from muzzle |

**Shipping and Storage Data:**

| | |
|---|---|
| Quantity-distance class/ SCG | 1.4S |
| Storage code | Class V |
| DOT shipping class | C |
| DOT designation | SMALL ARMS AMMUNITION |
| Drawing number | 7640995 |

**References:**

SB 700-20
TM 9-1005-206-14P/4
TM 9-1005-226-14
TM 9-1300-206

THIS PAGE INTENTIONALLY LEFT BLANK

## CARTRIDGE, CALIBER .22, BALL, HORNET, M65

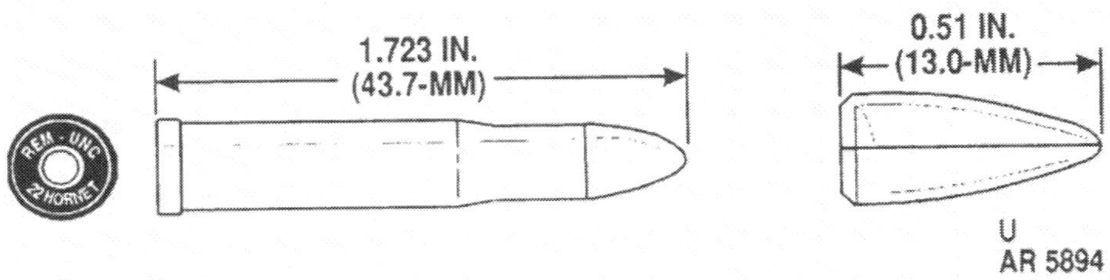


**Type Classification:**

OBS - MSR 11756003.

**Use:**

Caliber .22 Survival Weapons, M4 and M6. The cartridge is intended for use against small game.

**Description:**

BALL Cartridge.

**Tabulated Data:**

| | |
|---|---|
| DODAC | 1305-A088 |
| UNO serial number 0012 | |
| UNO proper shipping name | Cartridges for weapons, inert projectile |
| Weight | 103 gr |
| Length | 1.723 in. (43.7 mm) |
| Tracer | NA |
| Primer | Percussion |
| Fuze | NA |
| Explosive: | |
| Type | NA |
| Weight | NA |
| Incendiary: | |
| Type | NA |
| Weight | NA |
| Propellant: | |
| Type | Smokeless powder |
| Weight | 12.6 gr |

**Performance:**

| | |
|---|---|
| Chamber pressure | 43,000 psi |
| Velocity | 2750 fps, 15 ft from muzzle |

Shipping and Storage Data:

| | |
|---|---|
| Quantity-distance class/ SCG | 1.4S |
| Storage code | Class V |
| DOT shipping class | C |
| DOT designation | SMALL ARMS AMMUNITION |
| Drawing number | 7553664 |

**References:**

SB 700-20
TM 9-1005-206-14P/4
TM 9-1005-226-14
TM 9-1300-206

THIS PAGE INTENTIONALLY I,EFT BLANK

# CARTRIDGE, CALIBER .22, TRACER, LONG RIFLE, M861

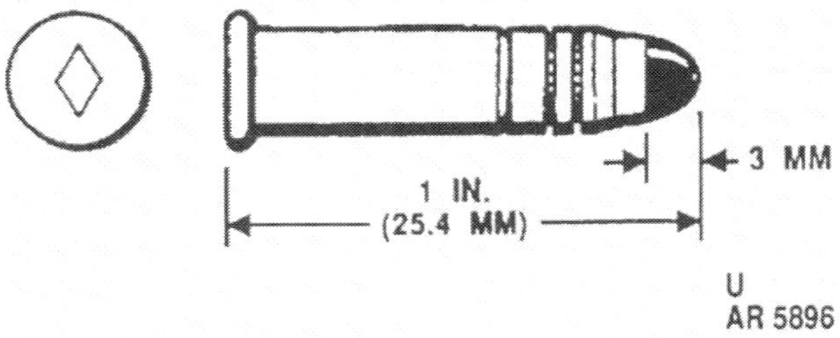


**Type Classification:**

STD - DA Letter.

**Use:**

Rifle M16A1 when adapted with the M261 Conversion Kit. The M861 cartridge, when fired from the rim-fired adapter M16A1 Rifle mounted on the Brewster Device, will increase the training proficiency in scaled range tank gunnery exercise by providing a visual display for tracking purposes.

**Description:**

TRACER Cartridge. The cartridge M861, .22 Caliber Long Rifle, rimfire with tracer is a commercially produced item, manufactured by Dynamite Nobel AG. The M861 is designed to be fired in the M16A1 rifle when adapted with the M261 Conversion Kit. It can be fired in the automatic or semi-automatic modes. The tracer (red) is visible between 27 and 192 feet in daylight to allow the gunner to observe and sense the impact of the tracer with normal vision and sighting devices when fired on the scaled range. The M861 is identified by a red painted tip.

**Tabulated Data:**

DODAC.................................... 1305-A090
UNO serial number 0012
UNO proper shipping
  name .................................... Cartridges for weapons, inert projectile
Weight ..................................... 48 gr'
Length...................................... 1 in. (25.4 mm)
Tracer ...................................... NA
Primer...................................... Percussion (Rimfire)
Fuze......................................... NA
Explosive:
  Type ..................................... NA
  Weight .................................. NA
Incendiary:
  Type ..................................... NA
  Weight .................................. NA
Propellant:
  Type ..................................... Smokeless powder
  Weight .................................. 2.5 gr

**Performance:**

Chamber pressure .................... 24,000 psi
Velocity.................................... - 990.75 fps, 302 m/s

**Shipping and Storage Data:**

Quantity-distance class/
  SCG...................................... 1.4S
Storage code............................ Class V
DOT shipping class - C
DOT designation ....................... SMALL ARMS AMMUNITION
Drawing number ....................... 9343015

THIS PAGE INTENTIONALLY LEFT BLANK

# CHAPTER 4

# CALIBER .30 CARBINE CARTRIDGES

THIS PAGE INTENTIONALLY LEFT BLANK

## CARTRIDGE, CALIBER .30, CARIBINE, BAI,L, M1

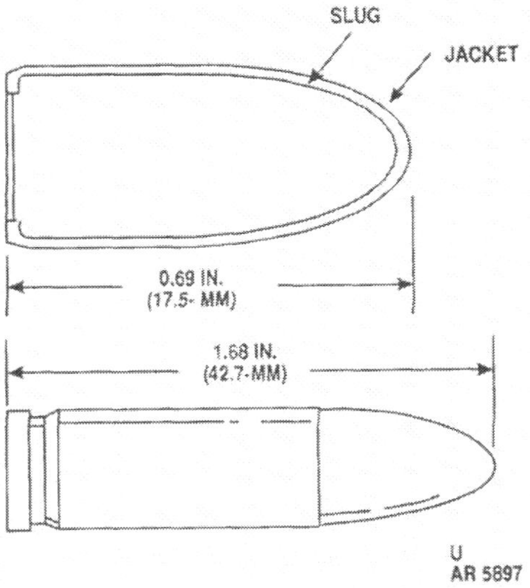


**Type Classification:**

OBS - MSR 11756003.

**Use:**

Caliber .30, Carbine, MI, M2 or M3. The cartridge is intended for use against personnel and unarmored targets.

**Description:**

BALL Cartridge. The cartridge is identified by the lack of bullet tip color code painting.

**Tabulated Data:**

DODAC .................................... 1305-A181
UNO serial number 0012
UNO proper shipping
name .................................... Cartridges for weapons, inert projectile
Weight .................................... 196 gr
Length .................................... 1.68 in. (42.7 mm)
Tracer .................................... NA
Primer .................................... Percussion
Fuze .................................... NA
Explosive:
Type .................................... NA
Weight .................................... NA
Incendiary:
Type .................................... NA
Weight .................................... NA
Propellant:
Type .................................... WC 820
Weight .................................... 13 gr
Projectile:
Weight .................................... 111 gr

**Performance:**

Chamber pressure 40,000 psi
Velocity .................................... 1900 fps, 53 ft from muzzle

**Shipping and Storage Data:**

Quantity-distance class/
SCG .................................... 1.4S
Storage code .................................... Class V
DOT shipping class - C
DOT designation .................................... SMALL ARMS AMMUNITION
Drawing number .................................... 6200954

**References:**

SB 700-20
TM 9-1300-206

THIS PAGE INTENTIONALLY LEFT BLANK

## CARTRIDG(E, CALIBER .30, CARBINES, GRENADE, M6

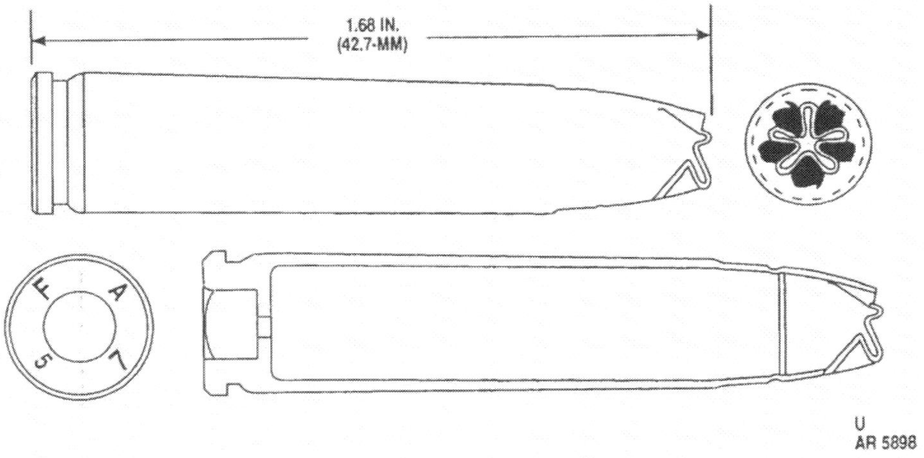


**Type Classification:**

OBS - MSR 11756003.

**Use:**

Caliber .30, Carbine, M1 or M2. The cartridge is currently used as a blank cartridge. 'The cartridge provides pressure to project grenades 1by using a grenade projection adapter.

**Description:**

GRENADE Cartridge. The cartridge identified by the rose-petal (rosette-crimp) closure of the cartridge case mouth.

**Tabulated Data:**

DODAC .................................... 1330-G837
UNO serial number .................. 0014
UNO proper shipping
  name .................................... Cartridges for weapons, blank
Weight ..................................... 103 gr
Length...................................... 1.68 in. (42.7 min)
Tracer ...................................... NA
Primer...................................... Percussion
Fuze......................................... NA
Explosive:
  Type ..................................... NA
  Weight .................................. NA
Incendiary:
  Type ..................................... NA
  Weight .................................. NA
Propellant:
  Type ..................................... IMR 480) and black powder
Weight ..................................... 21 gr

**Performance:**

Chamber pressure NA
Velocity ................................... 145 fps, 5 ft from muzzle of projectile grenade

**Shipping and Storage::**

Quantity-distance class/
  SCG...................................... 1.4S
Storage code............................ Class V
DO'I' shipping class - C
DOT designation ...................... SMALL ARMS AMMUNI'TION
Drawing number ...................... 6181079

**References:**

SB 700-20
TM 9-1300-206

THIS PAGE INTENTIONALLY LEFT BLANK

## CARTRIDGE, DUMMY, CALIBER .30, CARBINE, M13

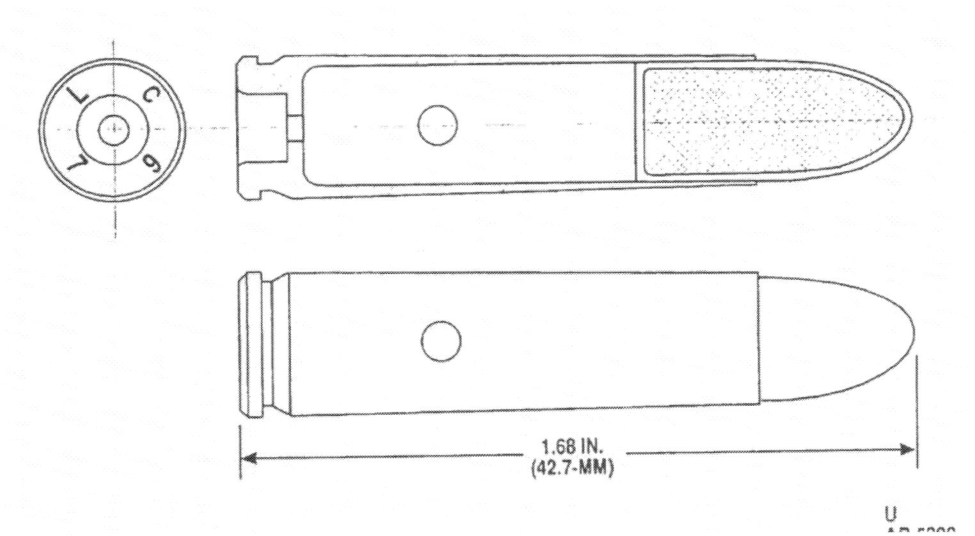


**Type Classification:**

OBS - MSR 11756003.

**Use:**

Caliber .30, Carbine, M1 or M2. The cartridge is completely inert and is used for practice in loading weapons, for simulated firing to detect flinching of personnel when firing, and for inspecting and testing the mechanisms of the weapon.

**Description:**

DUMMY Cartridge. The cartridge is identified by two holes in the cartridge case and the absence of the primer.

**Tabulated Data:**

DODAC.................................... 1305-A186
UNO serial number .................. 0339
UNO proper shipping
  name .................................... Cartridges for weapons, inert projectile
Weight ..................................... 177 gr
Length....................................... 1.68 in. (42.7 mm)
Tracer ....................................... NA
Primer....................................... NA
Fuze.......................................... NA
Explosive:
  Type ..................................... NA
  Weight .................................. - NA
Incendiary:
  Type ..................................... NA
  Weight .................................. NA
Propellant:
  Type ..................................... NA
  Weight .................................. NA

**Performance:**

Chamber pressure NA
Velocity .................................... NA

**Shipping and Storage Data:**

Quantity-distance class/
  SCG...................................... NA
Storage code............................. NA
DOT shipping class - NA
DOT designation ....................... NONEXPLOSIVE AMMUNITION
Drawing number ....................... 6177513

**References:**

SB 700-20
TM 9-1300-206

THIS PAGE INTENTIONALLY LEFT BLANK

## CARTRIDGE, CALIBER .30, CARBINE, BALL, HIGH PRESSURE TEST, M18

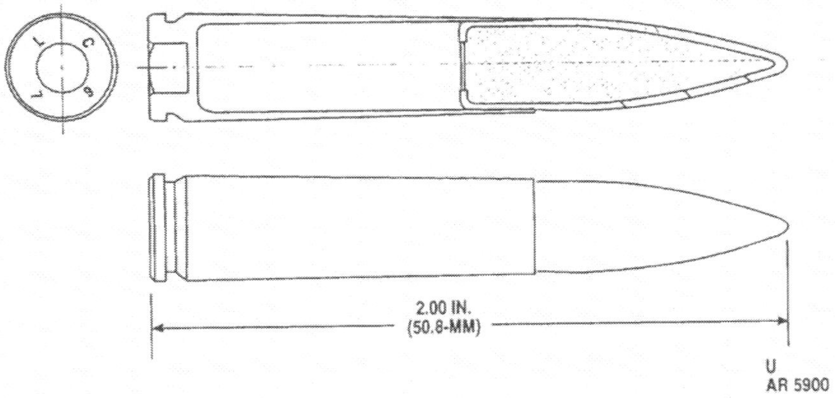


**Type Classification:**

STD - OTCM 36841.

**Use:**

Caliber .30, Carbine, M1 and M2. The cartridge is used for proof firing of caliber .30 carbine weapons during manufacture, test, or repair.

**Description:**

BALL Cartridge. The cartridge is identified by the stannic-stained (silvered) cartridge case.

**Tabulated Data:**

| | |
|---|---|
| DODAC | 1305-A185 |
| UNO serial number | 0012 |
| UNO proper shipping name | Cartridges for weapons, inert projectile |
| Weight | 243 gr |
| Length | 2 in. (50.8 mm) |
| Tracer | NA |
| Primer | Percussion |
| Fuze | NA |

| | |
|---|---|
| Explosive: | |
| Type | NA |
| Weight | NA |
| Incendiary: | |
| Type | NA |
| Weight | NA |
| Propellant: | |
| Type | HPC 5 |
| Weight | 14 gr |

**Performance:**

Chamber pressure 47,500 psi

| | |
|---|---|
| Velocity | NA |

**Shipping and Storage Data:**

| | |
|---|---|
| Quantity-distance class/ SCG | 1.4S |
| Storage code | Class V |
| DOT shipping class - C | |
| DOT designation | SMALL ARMS AMMUNITION |
| Drawing number | 6176267 |

**References:**

SB 700-20
TM 9-1300-206

THIS PAGE INTENTIONALLY LEFT BLANK

## CARTRIDGE, CALIBER .30, CARBINE, TRACER, M27

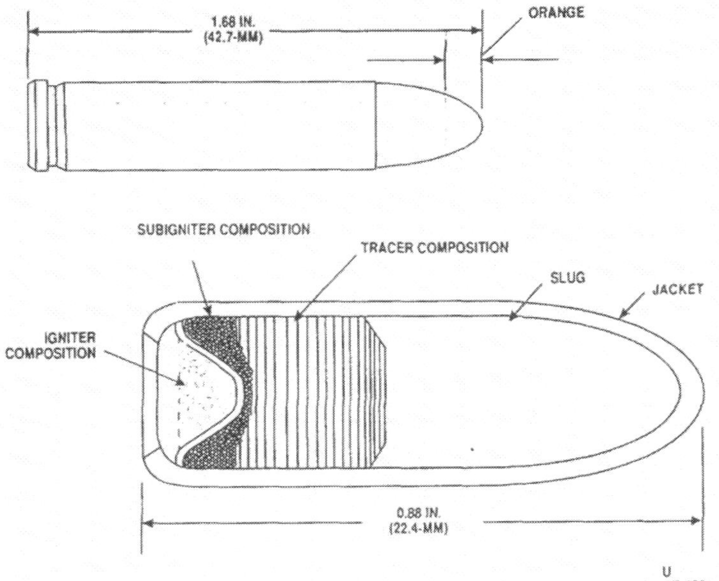


**Type Classification:**

OBS - MSR 11756003.

**Use:**

Caliber .30, Carbine, M1 and M2.

**Description:**

TRACER Cartridge. The cartridge is identified by an orange bullet tip.

**Function:**

In flight, the bullet exhibits a visible trace of full luminosity from a point not greater than 100 yards from the muzzle of the weapon to a point not less than 400 yards from the muzzle.

**Tabulated Data:**

DODAC.................................... 1305-A183
UNO serial number 0012
UNO proper shipping
name .................................... Cartridges for weapons, inert projectile

Weight .................................... 191 gr
Length.................................... 1.68 in. (42.7 mm)
Tracer .................................... NA
Primer.................................... Percussion
Fuze....................................... NA
Explosive:
Type .................................... NA
Weight ................................. NA
Incendiary:
Type .................................... NA
Weight ................................. NA
Propellant:
Type .................................... WC 820
Weight ................................. 13 gr

**Performance:**

Chamber pressure 40,000 psi
Velocity .................................. 1800 fps, 53 ft from muzzle

**Shipping and Storage Data:**

Quantity-distance class/
SCG.................................... 1.4S
Storage code........................... Class V

DOT shipping class ..................C
DOT designation ......................SMALL ARMS AMMUNITION
Drawing number........................76409309

**References:**

SB 700-20
TM 9-1300-206

# CHAPTER 5

# CALIBER .30 RIFLE AND MACHINE GUN CARTRIDGES

THIS PAGE INTENTIONALLY LEFT BLANK

**CARTRIDGE, CALIBER .30, BALL, HIGH PRESSURE TEST; MI**

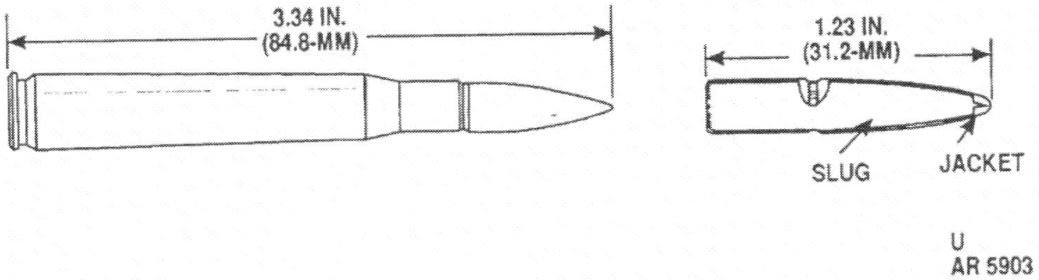


**Type Classification:**

STD - OTCM 37119.

**Use:**

All caliber .30 weapons except carbines. The cartridge is used to proof test caliber .30 rifles and machine guns during manufacture, test, or repair.

**Description:**

BALL Cartridge. The cartridge is identified by stannic-stained (silvered) cartridge case.

**Tabulated Data:**

| | |
|---|---|
| DODAC | 1305-A237 |
| UNO serial number 0012 | |
| UNO proper shipping name | Cartridges for weapons, inert projectile |
| Weight | 432 gr |
| Length | 3.34 in. (84.8 mm) |
| Tracer | NA |
| Primer | Percussion |
| Fuze | NA |
| Explosive: | |
| Type | NA |
| Weight | NA |
| Incendiary: | |
| Type | NA |
| Weight | NA |
| Propellant: | |
| Type | IMR 4198 |
| Weight | 52 gr |

**Performance:**

Chamber pressure 67,500 psi
Velocity ..................................... - NA

**Shipping and Storage Data:**

| | |
|---|---|
| Quantity-distance class/ SCG | 1.4S |
| Storage code | Class V |
| DOT shipping class | C |
| DOT designation | SMALL ARMS AMMUNITION |
| Drawing number | 6016308 |

**References:**

SB 700-20
TM 9-1005-222-12P/1
TM 9-1300-206

THIS PAGE INTENTIONALLY LEFT BLANK

## CARTRIDGE, CALIBER .30, TRACER, MI

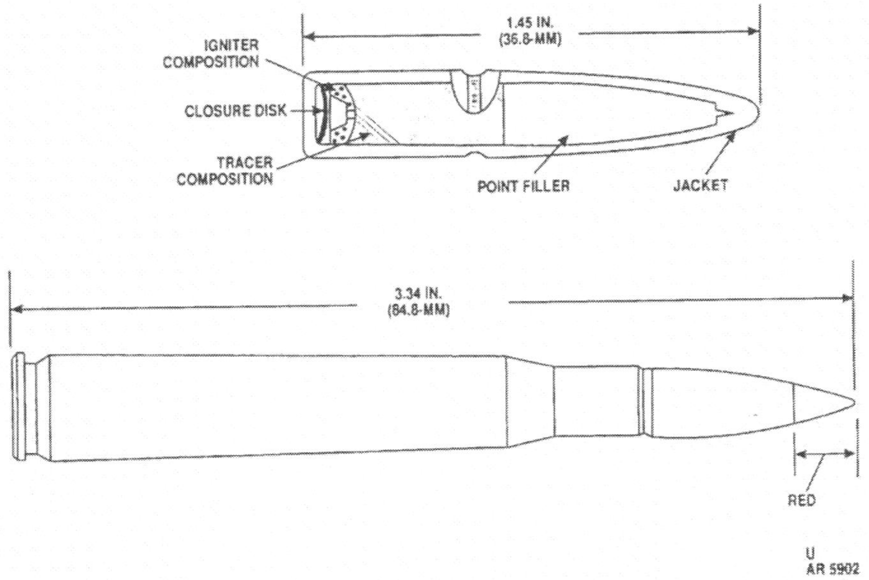


**Type Classification:**

OBS - MSR 11756003.

**Use:**

Machine Guns, Caliber .30, M37, M1919A4 and M1919A6; and Rifle, Caliber .30, M1. The tracer element is intended to permit visible observation of the bullet's in-flight path, or trajectory, to the target.

**Description:**

TRACER Cartridge. The cartridge is identified by a red bullet tip.

**Tabulated Data:**

| | |
|---|---|
| DODAC | 1305-A231 |
| UNO serial number 0012 | |
| UNO proper shipping name | Cartridges for weapons, inert projectile |
| Weight | 399 gr |
| Length | 3.34 in. (84.8 mm) |
| Tracer | R256 |
| Primer | Percussion |
| Fuze | NA |
| Explosive: | |
| Type | NA |
| Weight | NA |
| Incendiary: | |
| Type | NA |
| Weight | NA |
| Propellant: | |
| Type | IMR 4895 |
| Weight | 50 gr |

**Performance:**

| | |
|---|---|
| Chamber pressure | 52,000 psi |
| Velocity | 2665 fps, 78 ft from muzzle |

**Shipping and Storage Data:**

| | |
|---|---|
| Quantity-distance class/ SCG | 1.4S |
| Storage code | Class V |

DOT shipping class ..................C
DOT designation ........................SMALL ARMS AMMUNITION
Drawing number........................6006764

**References:**

SB 700-20
TM 9-1005-222-12P/1
TM 9-1300-206

## CARTRIDGE, CALIBER .30, ARMOR PIERCING, M2

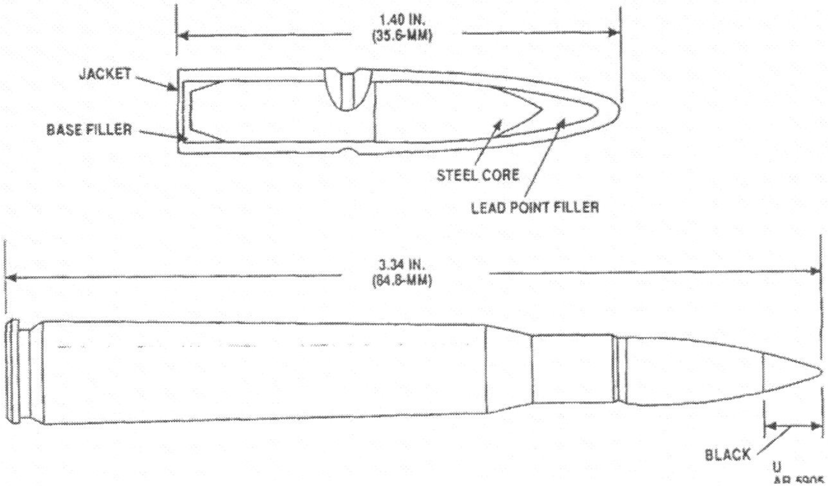


**Type Classification:**

OBS - MSR 11756003.

**Use:**

Machine Guns, Caliber .30, M37, M1919A4 and M1919A6; and Rifle, Caliber .30, M1.

**Description:**

ARMOR PIERCING Cartridge. The cartridge is identified by a black bullet tip.

**Function:**

Penetration, fired at 7/8-inch thick homogeneous armor plate at 100 yards, will be not less than 0.42 inch.

**Tabulated Data:**

| | |
|---|---|
| DODAC | 1305-A202 |
| UNO serial number | 0012 |
| UNO proper shipping name | Cartridges for weapons, inert projectile |
| Weight | 424 gr |
| Length | 3.34 in. (84.8 mm) |
| Tracer | NA |
| Primer | Percussion |
| Fuze | NA |
| Explosive: | |
| Type | NA |
| Weight | NA |
| Incendiary: | |
| Type | NA |
| Weight | NA |
| Propellant: | |
| Type | WC 852 |
| Weight | 55 gr |
| Projectile: | |
| Weight | 165.7 gr |

**Performance:**

| | |
|---|---|
| Chamber pressure | 54,000 psi |
| Velocity | 2715 fps, 78 ft from muzzle |

**Shipping and Storage Data:**

| | |
|---|---|
| Quantity-distance class/ SCG | 1.4S |
| Storage code | Class V |
| DOT shipping class | C |
| DOT designation | SMALL ARMS AMMUNITION |
| Drawing number | 6138194 |

**References:**

SB 700-20
TM 1005-222-12P/1
TM 9-1300-206

THIS PAGE INTENTIONALLY LEFT BLANK

## CARTRIDGE, CALIBER .30, BALL, M2

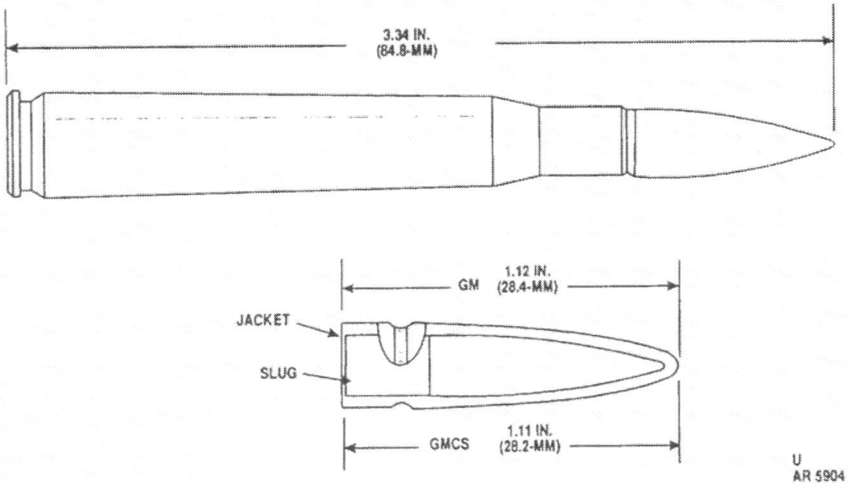


**Type Classification:**

OBS - MSR 11756003.

**Use:**

Machine Guns, Caliber .30, M37, M1919A4 and M1919A6; and Rifle, Caliber .30, M1. The cartridge is intended for use against personnel or unarmored targets.

**Description:**

BALL Cartridge. The cartridge is identified by a plain bullet tip.

**Tabulated Data:**

DODAC ............................................. 1305-A212
UNO serial number ........................... 0012
UNO proper shipping
name .......................................... Cartridges for weapons, inert projectile
Weight .............................................. 416 gr
Length ............................................... 3.34 in. (84.8 mm)
Tracer ............................................... NA
Primer ............................................... Percussion
Fuze .................................................. NA

Explosive:
Type ............................................ NA
Weight ......................................... NA
Incendiary:
Type ............................................ NA
Weight ......................................... NA
Propellant:
Type ............................................ IMR 4895
Weight ......................................... 50 gr

**Performance:**

Chamber pressure .............................. 50,000 psi
Velocity ............................................. 2740 fps, 78 ft from muzzle

**Shipping and Storage Data:**

Quantity-distance class/
SCG ........................................... 1.4S
Storage code ..................................... Class V
DOT shipping class ............................ C
DOT designation ................................ SMALL ARMS AMMUNITION
Drawing number ................................ 6137544

**References:**

SB 700-20
TM 9-1005-222-12P/1
TM 9-1300-206

THIS PAGE INTENTIONALLY LEFT BLANK

## CARTRIDGE, CALIBER.30, GRENADE, M3

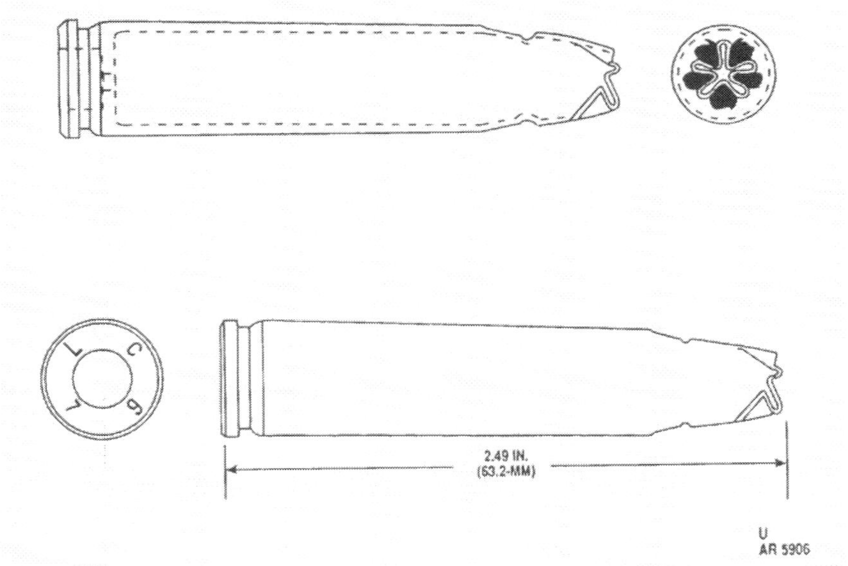


**Type Classification:**

OBS - MSR 11756003.

**Use:**

Rifle, Caliber .30, M1.

**Description:**

GRENADE Cartridge. The cartridge is identified by a rose-petal (rosette-crimped) closure of' the cartridge case mouth.

**Function:**

The cartridge provides the pressure, on functioning, to project rifle grenades when using a grenade projection adapter.

**Tabulated Data:**

| | |
|---|---|
| DODAC | 1330-G838 |
| UNO serial number | 0014 |
| UNO proper shipping name | Cartridges for weapons, blank |
| Weight | 248 gr |
| Length | 2.49 in. (63.2 mm) |
| Tracer- | NA |
| Primer | Percussion |
| Fuze | NA |
| Explosive: | |
| Type | NA |
| Weight | NA |
| Incendiary: | |
| Type | NA |
| Weight | NA |
| Propellant: | |
| Type | IMR 489)5 |
| Weight | 45 gr. |

**Performance:**

| | |
|---|---|
| Chamber pressure | NA |
| Velocity | 180 fps, 5.5 ft from muzzle/projected grenade |

Shipping and Storage Data:

| | |
|---|---|
| Quantity-distance class/ SCG | 1.4S |
| Storage code | Class V |

DOT shipping class ............................ C
DOT designation ................................ SMALL ARMS AMMUNITION
Drawing number................................. 6173949

**References:**
SB 700-20
TM 9-1005-222-12P/1
TM 9-1300-206

## CARTRIDGE, CALIBER .30, ARMOR PIERCING INCENDIARY, M14

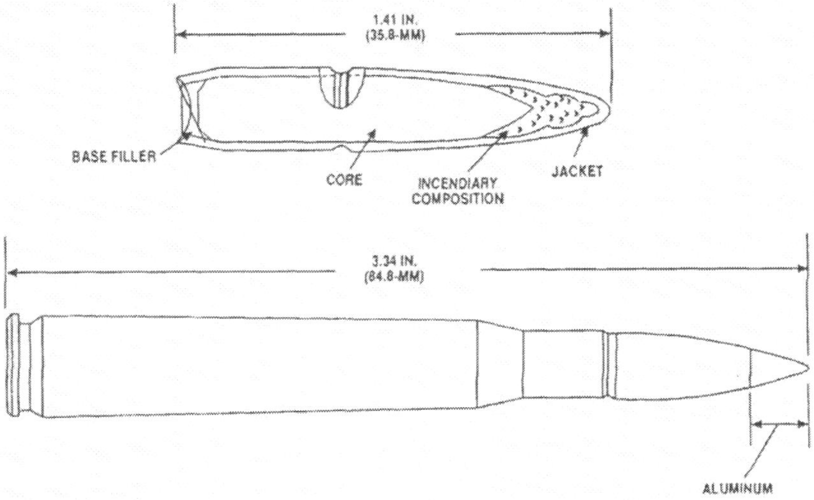


U
AR 5907

**Type Classification:**
OBS - MSR 11756003.

**Use:**

Machine Guns, Caliber .30, M37, M1919A4 and M1919A6; and Rifle, Caliber .30, M1.

**Description:**

ARMOR PIERCING INCENDIARY Cartridge. The cartridge is identified by an aluminum bullet tip.

**FUNCTION:**

Penetration, fired at a 7/8-inch steel plate at 100 yards, will not be less than 0.42 inch. Upon impact with a target, the incendiary composition bursts into flame and will ignite flammable material.

**Tabulated Data:**

DODAC ........................................... 1305-A202
UNO serial number ........................... 0012
UNO proper shipping
name ........................................ Cartridges for weapons, inert projectile
Weight ........................................... 407 gr
Length ........................................... 3.34 in. (84.8 mm)
Tracer ........................................... NA
Primer ........................................... Percussion
Fuze ............................................. NA
Explosive:
Type ........................................ NA
Weight ..................................... NA
Incendiary:
Type ........................................ IM 11
Weight ..................................... 2 gr
Propellant:
Type ........................................ WC 852
Weight ..................................... 50 gr

**Performance**

Chamber pressure .......................... 54,00() psi
Velocity .......................................... 2780 fps, 78 ft from muzzle

**Shipping and Storage Data:**

Quantity-distance class/
SCG ............................................. 1.4S
Storage code..................................... Class V

DOT shipping class ............................ C
DOT designation ................................ SMALL ARMS AMMUNITION
Drawing number................................. 7638431

**References:**

SB 700-20
TM 9-1005-222-12P/1
TM 9-1300-206

## CARTRIDGE, CALIBER .30, BALL, FRANGIBLE, M22

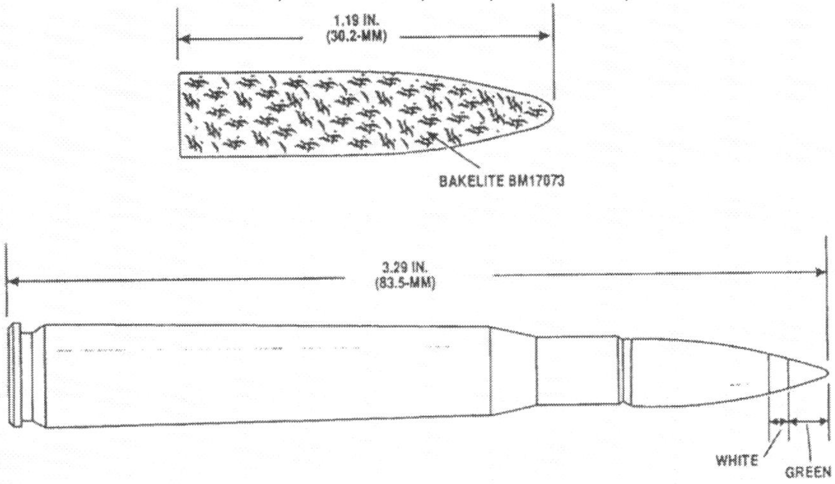


U
AR 5908

**Type Classification:**

OBS - MSR 11756003.

**Use:**

Machine Guns, Caliber .30, M37 and M1919A4; and Rifle, Caliber .30, M1.

**Description:**

BALL Cartridge. The cartridge is identified by a green bullet tip with a white ring to the rear of the green color.

**Function:**

The bullet disintegrates upon striking a hard or armored target, leaving a pencil-like mark to indicate a hit during gunnery practice.

**Tabulated Data:**

| | |
|---|---|
| DODAC | 1305-A220 |
| UNO serial number | 0012 |
| UNO proper shipping name | Cartridges for weapons, inert projectile |
| Weight | 320 gr |
| Length | 3.29 in. (83.5 mm) |
| Tracer | NA |
| Primer | Percussion |
| Fuze | NA |
| Explosive: | |
| Type | NA |
| Weight | NA |
| Incendiary: | |
| Type | NA |
| Weight | NA |
| Propellant: | |
| Type | SR 4759 |
| Weight | 11 gr |

**Performance:**

| | |
|---|---|
| Chamber pressure | NA |
| Velocity | 1320 fps, 53 ft from muzzle |

**Shipping and Storage Data:**

| | |
|---|---|
| Quantity-distance class/ SCG | 1.4S |
| Storage code | Class V |
| DOT shipping class | C |
| DOT designation | SMALL ARMS AMMUNITION |
| Drawing number | 7640951 |

**References:**

SB 700-20
TM 9-1300-206

THIS PAGE INTENTIONALLY LEFT BLANK

**CARTRIDGE, CALIBER .30, TRACER, M25**

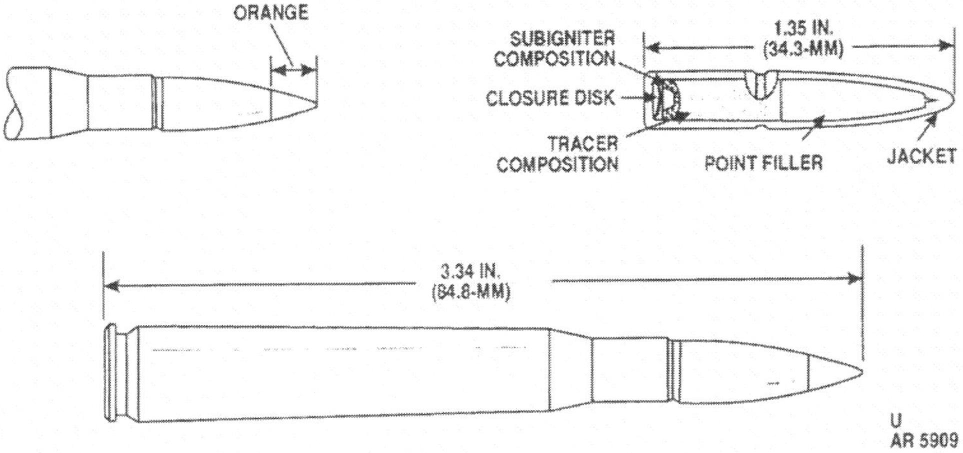


**Type Classification:**

OBS - MSR 11756003.
Use:

Machine Guns, Caliber .30, M37, M1917A1, M1919A4, and M1919A6; and Rifle, Caliber .30, M1. The cartridge is intended for use against personnel and unanmored targets.

**Description:**

TRACER Cartridge. The cartridge is identified by an orange bullet tip.

**Function:**

In flight, the bullet exhibits a trace of full luminosity from a point not greater than 75 yards from the muzzle of the weapon to a point not less than 900 yards from the muzzle.

**Tabulated Data:**

DODAC ........ 1305-A230
UNO serial number ........ 0012
UNO proper shipping
name ........ Cartridges for weapons, inert projectile
Weight ........ 401 gr
Length ........ 3.34 in. (84.8 mm)
Tracer ........ R321
Primer ........ Percussion
Fuze ........ NA

Explosive:
Type ........ NA
Weight ........ NA
Incendiary:
Type ........ NA
Weight ........ NA
Propellant:
Type ........ WC 852
Weight ........ 50 gr
Projectile:
Weight ........ 145.5 gr
Performance:

Chamber pressure ........ 50,000 psi
Velocity ........ 2665 fps, 78 ft from muzzle

**Shipping and Storage Data:**

Quantity-distance class/
SCG ........ 1.4S
Storage code ........ Class V
DOT shipping class ........ C
DOT designation ........ SMALL ARMS AMMUNITION
Drawing number ........ 7640667 (brass case); 7553949 (steel case)

**References:**

SB 700-20
TM 9-1005-222-12P/1
TM 9-1300-206

THIS PAGE INTENTIONALLY LEFT BLANK

## CARTRIDGE, DUMMY, CALIBER .30, M40

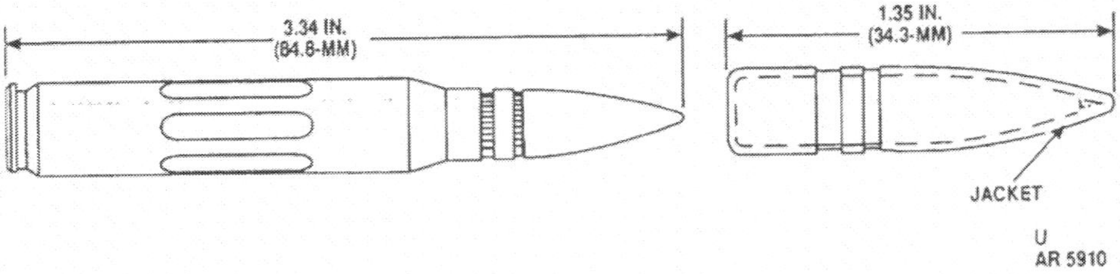


### Type Classification:

STD - OTCM 37119.

### Use:

All caliber .30 weapons except carbine. The cartridge is inert, and is used for practice in loading weapons, for simulated firing to detect flinching of personnel when firing, and for inspecting and testing the mechanisms of small arms weapons.

### Description:

DUMMY Cartridge. The cartridge is identified by six 1-inch longitudinal corrugations (flutings) in the cartridge case. In addition, there is no primer.

### Tabulated Data:

DODAC.......................................... 1305-A226
Weight.......................................... 268 gr
Length.......................................... 3.34 in. (84.8 mm)
Tracer.......................................... NA
Primer.......................................... NA
Fuze.......................................... NA

Explosive:
Type.......................................... NA
Weight.......................................... NA
Incendiary:
Type.......................................... NA
Weight.......................................... NA
Propellant:
Type.......................................... NA
Weight.......................................... NA

### Performance:

Chamber pressure.......................................... NA
Velocity.......................................... NA
Shipping and Storage Data:

Quantity-distance class/
SCG.......................................... NA
Storage code.......................................... NA
DOT shipping class.......................................... NA
DOT designation.......................................... NONEXPLOSIVE AMMUNITION
Drawing number.......................................... 7553591

### References:

SB 700-20
TM 9-1005-222-12P/1
TM 9-1300-206

THIS PAGE INTENTIONALLY LEFT BLANK

## CARTRIDGE, CALIBER .30, BALL, MATCH, M72

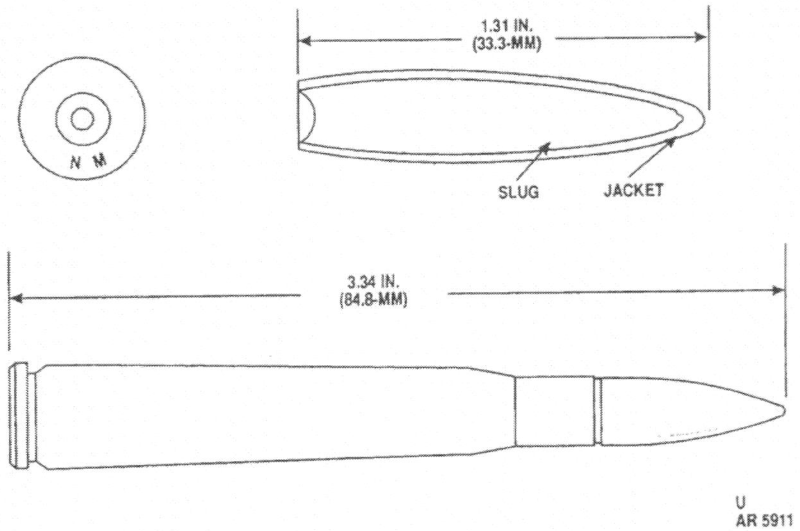


### Type Classification:

STD - OTCM 37119.

### Use:

Rifle, Caliber .30, M1; and Rifle, Caliber .30, National Match. The cartridge is designed to be fired only in weapons designated as competition rifles.

### Description:

BALL Cartridge. The cartridge is identified by head stampings on the base of the cartridge case with either the initials NM (National Match) or the word MATCH. The primer is uncrimped, the bullet has no crimped cannelure, and the case is not crimped to the bullet.

### Tabulated Data:

| | |
|---|---|
| DODAC | 1305-A247 |
| UNO serial number | 0012 |
| UNO proper shipping name | Cartridges for weapons, inert projectile |
| Weight | 425 gr |
| Length | 3.34 in. (84.8 mm) |
| Tracer | NA |
| Primer | Percussion |
| Fuze | NA |
| Explosive: | |
| Type | NA |
| Weight | NA |
| Incendiary: | |
| Type | NA |
| Weight | NA |
| Propellant: | |
| Type | IMR 4895 |
| Weight | 50 gr |
| Performance: | |
| Chamber pressure | 50,000 psi |
| Velocity | 2640 fps, 78 ft from muzzle |

### Shipping and Storage Data:

| | |
|---|---|
| Quantity-distance class/ SCG | 1.4S |
| Storage code | Class V |
| DOT shipping class | C |
| DOT designation | SMALL ARMS AMMUNITION |
| Drawing number | 8595432 |

### References:

SB 700-20
TM 9-1005-222-12P/1
TM 9-1300-206

THIS PAGE INTENTIONALLY LEFT BLANK

## CARTRIDGE, CALIBER .30, BLANK, M1909

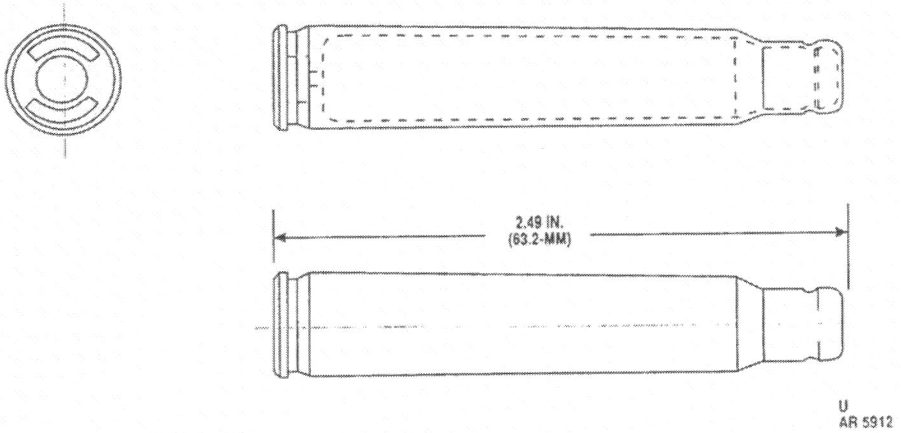


**Type Classification:**

OBS - MSR 11756003.

**Use:**

Machine Guns, Caliber .30, M1919A4 and M1919A6; and Rifle, Caliber .30, MI. The cartridge is used for simulated firing in training or for saluting purposes.

**Description:**

BLANK Cartridge. The cartridge is identified by the absence of a bullet and has a crimped cartridge case mouth.

**Tabulated Data:**

DODAC ........ 1305-A222
UNO serial number ........ 0014
UNO proper shipping
name ........ Cartridges for weapons, blank
Weight ........ 218 gr
Length ........ 2.49 in. (63.2 mm)
Tracer ........ NA
Primer ........ Percussion
Fuze ........ NA
Explosive:
Type ........ NA
Weight ........ NA
Incendiary:
Type ........ NA
Weight ........ NA
Propellant:
Type ........ SR 4990
Weight ........ 12 gr
Performance:
Chamber pressure ........ NA
Velocity ........ NA

**Shipping and Storage Data:**

Quantity-distance class/
SCG ........ 1.4S
Storage code ........ Class V
DOT shipping class ........ C
DOT designation ........ SMALL ARMS AMMUNITION
Drawing number ........ 6006152

**References:**

SB 700-20
TM 9-1005-222-12P/1
TM 9-1300-206

THIS PAGE INTENTIONAILY LEFT BLANK

# CHAPTER 6

# CALIBER .32 CARTRIDGES

THIS PAGE INTENTIONALLY LEFT BLANK

## CARTRIDGE, CALIBER .32, BALL

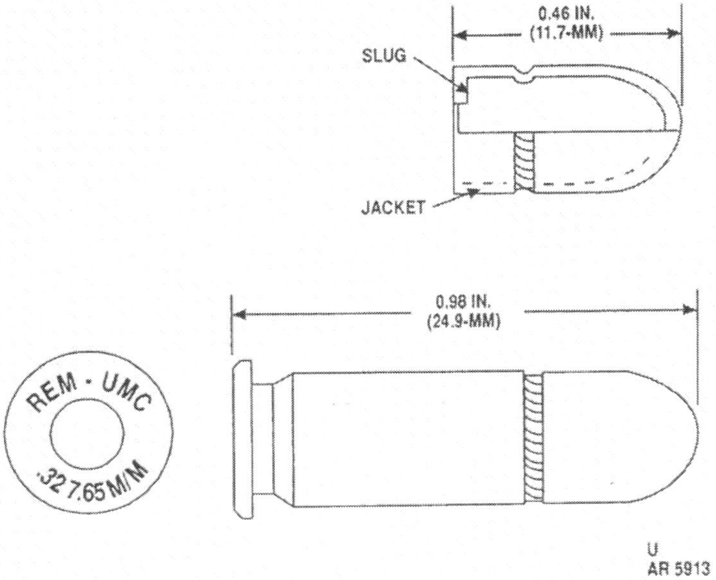


**Type Classification:**

Commercial.

**Use:**

Caliber .32 weapons. The cartridge is used in caliber .32 automatic pistols for defensive use against personnel.

**Description:**

BALL Cartridge.

**Tabulated Data:**

| | |
|---|---|
| DODAC | 1305-A350 |
| UNO serial number | 0012 |
| UNO proper shipping name | Cartridges for weapons, inert projectile |
| Weight | 115 gr |
| Length | 0.98 in. (24.9 mm) |
| Tracer | NA |
| Primer | Percussion |
| Fuze | NA |
| Explosive: | |
| Type | NA |
| Weight | NA |
| Incendiary: | |
| Type | NA |
| Weight | NA |
| Propellant: | |
| Type | Smokeless powder |
| Weight | MBR |

Performance:

| | |
|---|---|
| Chamber pressure | 15,000 psi |
| Velocity | 900 fps, 15 ft from muzzle |

**Shipping and Storage Data:**

| | |
|---|---|
| Quantity-distance class/ SCG | 1.4S |
| Storage code | Class V |
| DOT shipping class | C |
| DOT designation | SMALL ARMS AMMUNITION |
| Drawing number | 10524631 |

References:

SB 700-20
TM 9-1005-206-14P/3
TM 9-1300-206

THIS PAGE INTENTIONALLY LEFT BLANK

# CHAPTER 7
# CALIBER.38 CARTRIDGES

THIS PAGE INTENTIONALLY LEFT BLANK

## CARTRIDGE, CALIBER .38, BALL

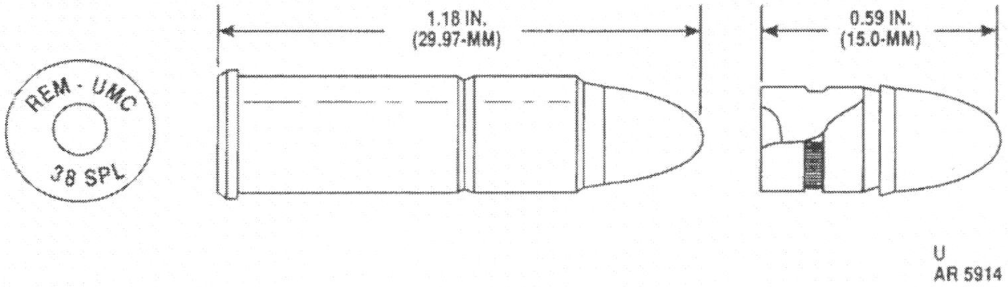


**Type Classification:**

OBS - MSR 11756003.

**Use:**

Caliber .38 weapons. The cartridge is for CONUS-guard or security use in caliber .38 weapons.

**Description:**

BALL Cartridge. The cartridge has an unjacketed lead bullet.

**Tabulated Data:**

| | |
|---|---|
| DODAC | 1305-A408 |
| UNO serial number | 0012 |
| UNO proper shipping name | Cartridges for weapons, inert projectile |
| Weight | 196 gr |
| Length | 1.18 in. (29.97 mm) |
| Tracer | NA |
| Primer | Percussion |
| Fuze | NA |
| Explosive: | |
| Type | NA |
| Weight | NA |
| Incendiary: | |
| Type | NA |
| Weight | NA |
| Propellant: | |
| Type | Smokeless powder |
| Weight | MBR |

**Performance:**

| | |
|---|---|
| Chamber pressure | 13,000 psi |
| Velocity | 725 fps, 15 ft from muzzle |

**Shipping and Storage Data:**

| | |
|---|---|
| Quantity-distance class/ SCG | 1.4S |
| Storage code | Class V |
| DOT shipping class | C |
| DOT designation | SMALL ARMS AMMUNITION |
| Drawing .number | 10524005 |

**References:**

SB 700-20
TM 9-1005-206-14&P-1
TM 9-1005-206-14P/3
TM 9-1300-206

THIS PAGE INTENTIONALLY

## CARTRIDGE, CALIBER .38, BALL, SPECIAL, M41

**Type Classification:**

OBS - MSR 08836015.

**Use:**

Revolver. The cartridge is intended for use in personal defense hand guns.

**Description:**

BALL Cartridge. The cartridge has a jacketed bullet.

**Tabulated Data:**

DODAC ........ 1305-A400
UNO serial number ........ 0012
UNO proper shipping name ........ Cartridges for weapons, insert projectile
Weight ........ 203 gr
Length ........ 1.55 in. (39.4 mm)
Tracer ........ NA
Primer ........ Percussion
Fuze ........ NA
Explosive:
- Type ........ NA
- Weight ........ NA

Incendiary:
- Type ........ NA
- Weight ........ NA

Propellant:
- Type ........ SR 7325
- Weight ........ 4.8 gr

**Performance:**

Chamber pressure ........ 16,000 psi
Velocity ........ 950 fps, 15 ft from muzzle

**Shipping and Storage Data:**

Quantity-distance class/SCG ........ 1.4S
Storage code ........ Class V
DOT shipping class ........ C
DOT designation ........ SMALL ARMS AMMUNITION
Drawing number ........ 7553580

**References:**

SB 700-20
TM 9-1005-206-14&P-1
TM 9-1005-206-14P/3
TM 9-1300-206

THIS PAGE INTENTIONALLY LEFT BLANK

## CARTRIDGE, CALIBER .38, WAD CUTTER

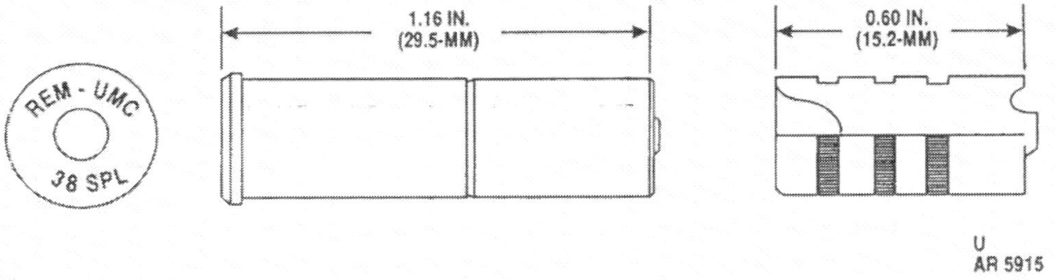


**Type Classification:**

Commercial.

**Use:**

Revolver, Special, S & W, K-38. The cartridge is used only in p stol competition firing and marksmanship training.

**Description:**

WAD CUTTER Cartridge.

**Function:**

The bullet is designed with a concave forward end to act as a paper punch cutting a clean hole in targets.

**Tabulated Data:**

| | |
|---|---|
| DODAC | 1305-A404 |
| UNO serial number | 0012 |
| UNO proper shipping name | Cartridges for weapons, inert projectile |
| Weight | 220 gr |
| Length | 1.16 in. (29.5 mm) |
| Tracer | NA |
| Primer | Percussion |
| Fuze | NA |
| Explosive: | |
| Type | NA |
| Weight | NA |
| Incendiary: | |
| Type | NA |
| Weight | NA |
| Propellant: | |
| Type | Smokeless powder |
| Weight | MBR |

**Performance:**

| | |
|---|---|
| Chamber pressure | 16,000 psi |
| Velocity | 800 fps, 15 ft from muzzle |

**Shipping and Storage Data:**

| | |
|---|---|
| Quantity-distance class/ SCG | 1.4S |
| Storage code | Class V |
| DOT shipping class | C |
| DOT designation | SMALL ARMS AMMUNITION |
| Drawing number | 10521459 |

**References:**

SB 700-20
TM 9-1005-206-14&P-1
TM 9-1005-206-14P/3
TM 9-1300-206

THIS PAGE INTENTIONALLY LEFT BLANK

# CHAPTER 8

# CALIBER .45 CARTRIDGES

THIS PAGE INTENTIONALLY LEFT BLANK

## CARTRIDGE, CALIBER .45, WAD CUTTER

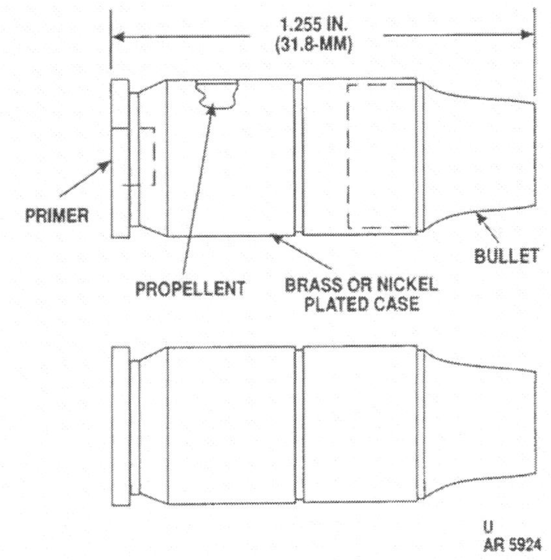


**Type Classification:**

OBS - MSR 08836015.

**Use:**

Pistol, Caliber .45, M1911A1; and Pistol, Caliber .45, National Match.

**Description:**

WAD CUTTER Cartridge. The cartridge is identified by the bullet shape which has the general appearance of a truncated cone.

**Function:**

The ogive of the bullet has a sharp shoulder that acts as a hole punch to cut a clean hole in the target.

**Tabulated Data:**

DODAC ------------------------------- 1305-A482
UNO serial number----------------- 0012
UNO proper shipping name------ Cartridges for weapons, inert projectile

Weight -------------------------------- 330 gr
Length -------------------------------- 1.255 in. (31.8 mm)
Tracer --------------------------------- NA
Primer--------------------------------- Percussion
Fuze ----------------------------------- NA
Explosive:
Type ---------------------------- NA
Weight-------------------------- NA
Incendiary:
Type ---------------------------- NA
Weight-------------------------- NA
Propellant:
Type ---------------------------- Smokeless powder
Weight-------------------------- MBR

**Performance:**

Chamber pressure ----------------- 18,000 psi
Velocity ------------------------------- 765 fps, 15 ft from muzzle

**Shipping and Storage Data:**

Quantity-distance class/
SCG----------------------------- 1.4S
Storage code------------------------ Class V

DOT shipping class ----------------C
DOT designation--------------------SMALL ARMS AMMUNITION
Drawing number --------------------10521458

**References:**

SB 700-20
TM 9-1005-211-12
TM 9-1300-206

## CARTRIDGE, CALIBER .45, BALL, HIGH PRESSURE TEST, M1

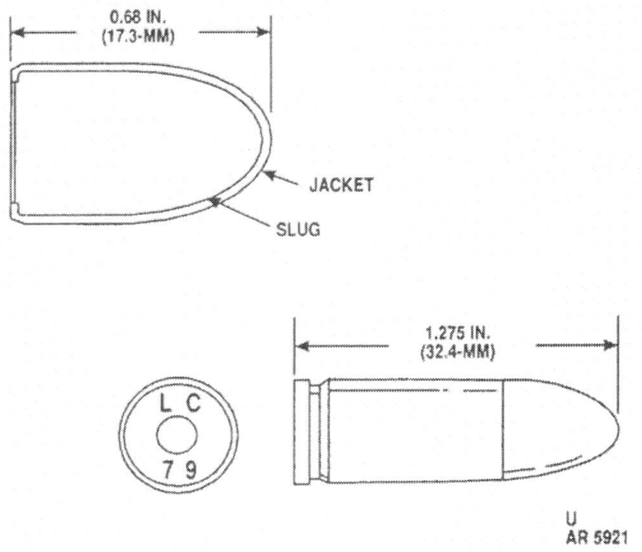


**Type Classification:**

STD - OTCM 36841.

**Use:**

Test of all caliber .45 weapons. The cartridge is for proof testing caliber .45 pistols and submachine guns during manufacture, test, or repair.

**Description:**

BALL Cartridge. The cartridge is identified by a stannic-stained (silvered) cartridge case.

**Tabulated Data:**

DODAC ---------------------------------1305-A480
UNO serial number------------------0012
UNO proper shipping name------Cartridges for weapons, inert projectile
Weight---------------------------------332 gr
Length ---------------------------------1.275 in. (32.4 mm)
Tracer----------------------------------NA
Primer ---------------------------------Percussion
Fuze------------------------------------NA
Explosive:
Type -----------------------------NA
Weight--------------------------- NA
Incendiary:
Type ----------------------------- NA
Weight--------------------------- NA
Propellant:
Type ----------------------------- SR 7970
Weight--------------------------- 7 gr

**Performance:**

Chamber pressure------------------ 22,000 psi
Velocity---------------------------------- NA

**Shipping and Storage Data:**

Quantity-distance class/
SCG----------------------------- 1.4S
Storage code------------------------ Class V
DOT shipping class----------------- C
DOT designation ------------------- SMALL ARMS AMMUNITION
Drawing number-------------------- C6000502

**References:**

SB 700-20
TM 9-1005-211-12
TM 9-1005-229-12
TM 9-1300-206

THIS PAGE INTENTIONALLY LEFT BLANK

## CARTRIDGE, CALIBER .45, BLANK, M9

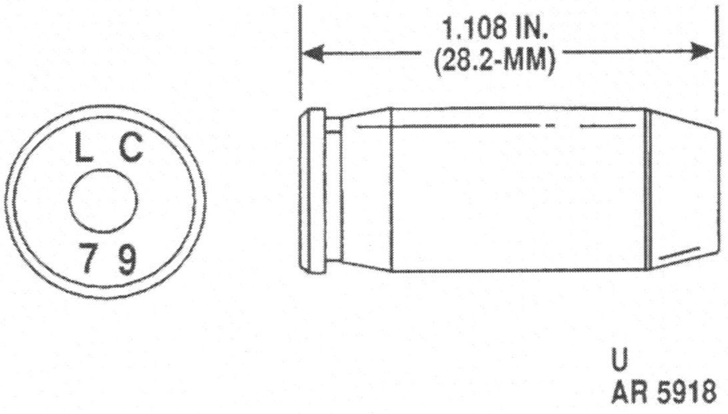


**Type Classification:**

STD - AMCTC 505.

**Use:**

Pistol, Automatic, Caliber .45, M1911A1. The cartridge is used for simulated firing in training maneuvers and for saluting purposes.

**Description:**

BLANK Cartridge. The cartridge is identified by the absence of a bullet and the tapered mouth of the cartridge case.

**Tabulated Data:**

| | |
|---|---|
| DODAC | 1305-A476 |
| UNO serial number | 0014 |
| UNO proper shipping name | Cartridges for weapons, blank |
| Weight | 104 gr |
| Length | 1.108 in. (28.2 mm) |
| Tracer | NA |
| Primer | Percussion |
| Fuze | NA |
| Explosive: | |
| Type | NA |
| Weight | NA |
| Incendiary: | |
| Type | NA |
| Weight | NA |
| Propellant: | |
| Type | SR 4990 |
| Weight | 10 gr |

**Performance:**

| | |
|---|---|
| Chamber pressure | NA |
| Velocity | NA |

**Shipping and Storage Data:**

| | |
|---|---|
| Quantity-distance class/ SCG | 1.4S |
| Storage code | Class V |
| DOT shipping class | C |
| DOT designation | SMALL ARMS AMMUNITION |
| Drawing number | 10523805 |

**References:**

SB 700-20
TM 9-1005-211-12
TM 9-1300-206

THIS PAGE INTENTIONALLY LEFT BLANK

## CARTRIDGE, CALIBER .45, TRACER, M26

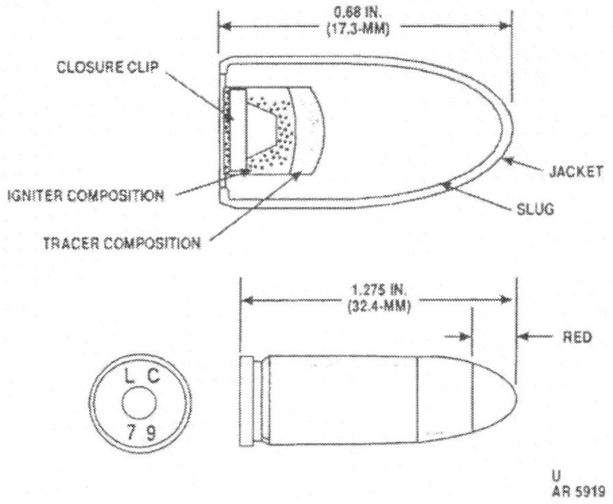


### Type Classification:

CON - MSR 11756003.

### Use:

Submachine Gun, Caliber .45, M3A1; and Pistol, Caliber .45, M1911A1. The tracer is intended to permit observation of the bullet's in-flight path or trajectory to the point of impact.

### Description:

TRACER Cartridge. The cartridge is identified by a red bullet tip.

### Tabulated Data:

| | |
|---|---|
| DODAC | 1305-A479 |
| UNO serial number | 0012 |
| UNO proper shipping name | Cartridges for weapons, inert projectile |
| Weight | 331 gr |
| Length | 1.275 in. (32.4 mm) |
| Tracer | R256 |
| Primer | Percussion |
| Fuze | NA |

| | |
|---|---|
| Explosive: | |
| Type | NA |
| Weight | NA |
| Incendiary: | |
| Type | NA |
| Weight | NA |
| Propellant: | |
| Type | SR 7970 |
| Weight | 6 gr |

### Performance:

| | |
|---|---|
| Chamber pressure | 19,000 psi |
| Velocity | 885 fps, 25.5 ft from muzzle |

### Shipping and Storage Data:

| | |
|---|---|
| Quantity-distance class/ SCG | 1.4S |
| Storage code | Class V |
| DOT shipping class | C |
| DOT designation | SMALL ARMS AMMUNITION |
| Drawing number | 10534359 |

### References:

SB 700-20
TM 9-1005-211-12
TM 9-1005-229-12
TM 9-1300-206

THIS PAGE INTENTIONALLY LEFT BLANK

## CARTRIDGE, CALIBER .45, LINE THROWING, M32

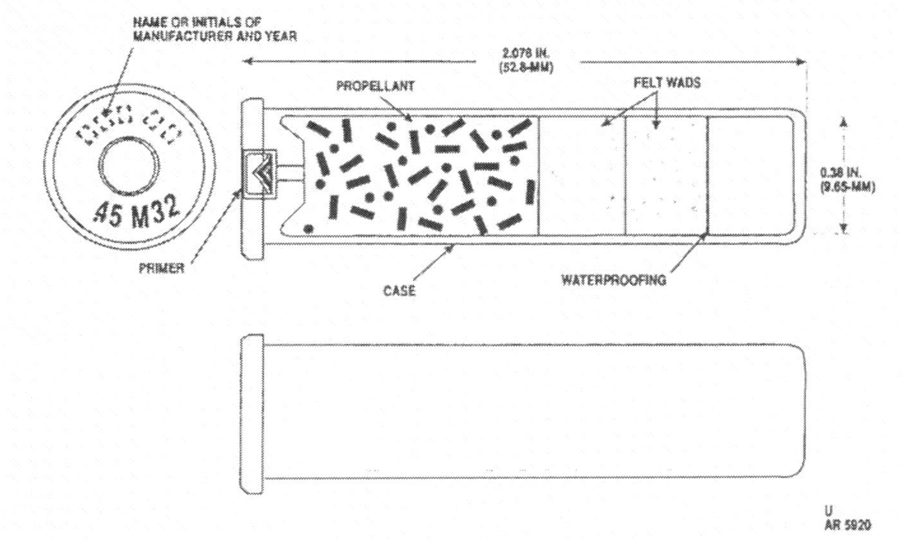


**Type Classification:**

STD A - OTCM 36841.

**Use:**

Line Throwing Gun, Caliber .45/.70.

**Description:**

LINE THROWING Cartridge. The cartridge is composed of a brass case, primer, commercial smokeless powder; and two felt wads coated on both sides with gummed paper. The wads are sealed with red lacquer. The cartridge case mouth is crimped to prevent entry of the line throwing rod.

**Function:**

The line throwing rod is placed in the weapon barrel. Cartridge initiation provides pressure to force the rod out of the barrel and propel the line to the target.

**Tabulated Data:**

DODAC ------------------------------- 1305-A477
UNO serial number----------------- 0014
UNO proper shipping
name--------------------------- Cartridges for weapons, blank
Weight -------------------------------- 192 gr
Length -------------------------------- 2.078 in. (52.8 mm)
Tracer--------------------------------- NA
Primer--------------------------------- Percussion
Fuze----------------------------------- NA
Explosive:
Type ---------------------------- NA
Weight-------------------------- NA
Incendiary:
Type ---------------------------- NA
Weight-------------------------- NA
Propellant:
Type ---------------------------- Smokeless powder
Weight-------------------------- MBR

**Performance:**

Chamber pressure----------------- 20,000 psi
Velocity ------------------------------- NA

**Shipping and Storage Data:**

Quantity-distance class/
SCG----------------------------- 1.4S
Storage code------------------------ Class V

DOT shipping class ----------------C
DOT designation--------------------SMALL ARMS AMMUNITION
Drawing number --------------------7553068

**References:**

SB 700-20
TM 9-1300-206

**CARTRIDGE, CALIBER .45, BALL, M1911**

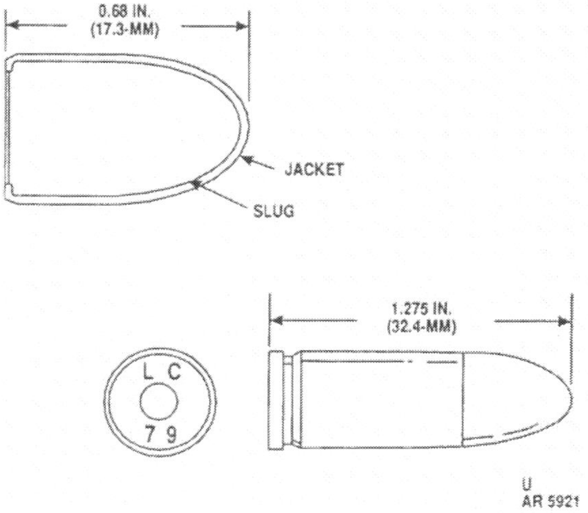


**Type Classification:**

STD - OTCM 36841.

**Use:**

Submachine Gun, Caliber .45, M3A1; and Pistol, Caliber .45, M1911A1. The cartridge is intended for use against personnel.

**Description:**

BALL Cartridge. The cartridge is identified by the plain bullet tip.

**Tabulated Data:**

DODAC ------------------------------- 1305-A475
UNO serial number----------------- 0012
UNO proper shipping
name ---------------------------- Cartridges for weapons, inert projectile
Weight--------------------------------- 331 gr
Length -------------------------------- 1.275 in. (32.4 mm)
Tracer--------------------------------- NA
Primer -------------------------------- Percussion
Fuze----------------------------------- NA

Explosive:
Type ---------------------------- NA
Weight-------------------------- NA
Incendiary:
Type ---------------------------- NA
Weight-------------------------- NA
Propellant:
Type ---------------------------- SR 7970
Weight-------------------------- 5 gr

**Performance:**

Chamber pressure----------------- 19,000 psi
Velocity ------------------------------ 885 ± 25 fps, 25.5 ft from muzzle

**Shipping and Storage Data:**

Quantity-distance class/
SCG---------------------------- 1.4S
Storage code----------------------- Class V
DOT shipping class---------------- C
DOT designation ------------------- SMALL ARMS AMMUNITION
Drawing number-------------------- 6000503

**References:**

SB 700-20
TM 9-1005-211-12
TM 9-1005-229-12
TM 9-1300-206

THIS PAGE INTENTIONALLY LEFT BLANK

## CARTRIDGE, CALIBER .45, BALL, MATCH, M1911

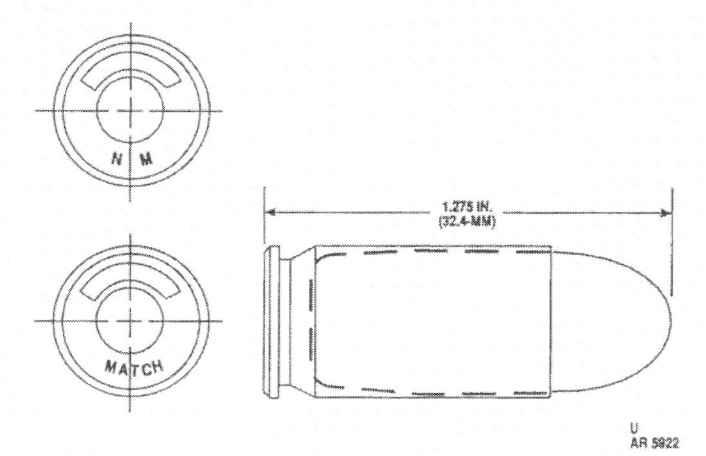


**Type Classification:**

STD - OTCM 36841.

**Use:**

Pistol, Caliber .45, M1911A1; and Pistol, Caliber .45, National Match. The cartridge is to be used only in those weapons designated as competition pistols.

**Description:**

BALL Cartridge. The cartridge is identified by head stampings, either NM (National Match) or MATCH, on the base of the cartridge case.

**Tabulated Data:**

| | |
|---|---|
| DODAC | 1305-A483 |
| UNO serial number | 0012 |
| UNO proper shipping name | Cartridges for weapons, inert projectile |
| Weight | 334 gr |
| Length | 1.275 in. (32.4 mm) |
| Tracer | NA |
| Primer | Percussion |
| Fuze | NA |
| Explosive: | |
| Type | NA |
| Weight | NA |
| Incendiary: | |
| Type | NA |
| Weight | NA |
| Propellant: | |
| Type | SR 7970 |
| Weight | 5 gr |

**Performance:**

| | |
|---|---|
| Chamber pressure | 19,000 psi |
| Velocity | 885 ± 25 fps, 25.5 ft from muzzle |

**Shipping and Storage Data:**

| | |
|---|---|
| Quantity-distance class/ SCG | 1.4S |
| Storage code | NA |
| DOT shipping class | C |
| DOT designation | SMALL ARMS AMMUNITION |
| Drawing number | 8596136 |

**References:**

SB 700-20
TM 9-1005-211-12
TM 9-1300-206

THIS PAGE INTENTIONALLY LEFT BLANK

## CARTRIDGE, DUMMY, CALIBER .45, M1921

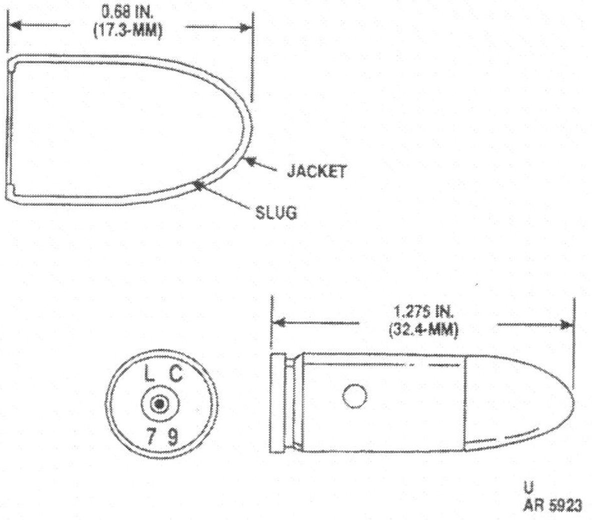


**Type Classification:**

STD - OTCM 36841.

**Use:**

Submachine Gun, Caliber .45, M3A1; and Pistol, Caliber .45, M1911A1. The cartridge is used during simulated firing exercises to detect flinching of personnel when firing, and to inspect and test the mechanism of the weapons.

**Description:**

DUMMY Cartridge. The cartridge is identified by the drilled holes in the cartridge case and the absence of a primer.

**Tabulated Data:**

DODAC ------------------------------- 1305-A501
Weight--------------------------------- 305 gr
Length -------------------------------- 1.275 in. (32.4 mm)
Tracer--------------------------------- NA
Primer -------------------------------- NA
Fuze----------------------------------- NA
Explosive:
Type ---------------------------- NA
Weight ------------------------- NA
Incendiary:
Type ---------------------------- NA
Weight------------------------- NA
Propellant:
Type ---------------------------- NA
Weight------------------------- NA

**Performance:**

Chamber pressure----------------- NA
Velocity------------------------------- NA

**Shipping and Storage Data:**

Quantity-distance class/
SCG ---------------------------------- NA
Storage code------------------------ NA
DOT shipping class--------------- NA
DOT designation ------------------ NONEXPLOSIVE AMMUNITION
Drawing number------------------- 6006253

**References:**

SB 700-20
TM 9-1005-211-12
TM 9-1005-229-12
TM 9-1300-206

THIS PAGE INTENTIONALLY LEFT BLANK

# CHAPTER 9

# CALIBER .50 CARTRIDGES

THIS PAGE INTENTIONALLY LEFT BLANK

## CARTRIDGE, CALIBER .50, BALL, HIGH PRESSURE TEST, M1

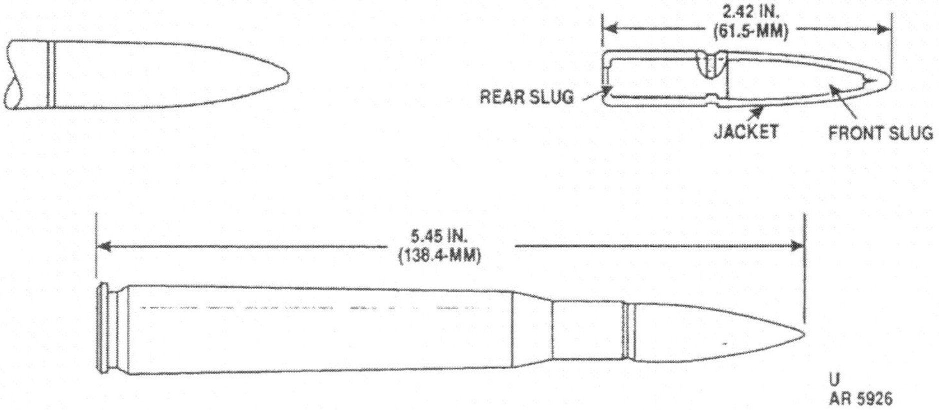


**Type Classification:**

STD - OTCM 36841.

**Use:**

All caliber .50 weapons except M8C Spotting Gun. The cartridge is intended for use in proof testing caliber .50 weapons (except the 106mm Spotting Rifle) during manufacture, test, or repair.

**Description:**

BALL Cartridge. The cartridge is identified by a stannic-stained (silvered) cartridge case.

**Tabulated Data:**

| | |
|---|---|
| DODAC | 1305-A575 |
| UNO serial number | 0339 |
| UNO proper shipping name | Cartridges for weapons, inert projectile |
| Weight | 2082 gr |
| Length | 5.45 in. (138.4 mm) |
| Tracer | NA |
| Primer | Percussion |
| Fuze | NA |

| | |
|---|---|
| Explosive: | |
| Type | NA |
| Weight | NA |
| Incendiary: | |
| Type | NA |
| Weight | NA |
| Propellant: | |
| Type | WC 860 |
| Weight | 240 gr |

**Performance:**

| | |
|---|---|
| Chamber pressure | 65,000 psi |
| Velocity | NA |

**Shipping and Storage Data:**

| | |
|---|---|
| Quantity-distance class/ SCG | 1.4S |
| Storage code | Class V |
| DOT shipping class | C |
| DOT designation | SMALL ARMS AMMUNITION |
| Drawing number | 5544097 |

**References:**

SB 700-20
TM 9-1005-213-10
TM 9-1005-231-10
TM 9-1300-206

THIS PAGE INTENTIONALLY LEFT BLANK

## CARTRIDGE, CALIBER .50, BLANK, M1

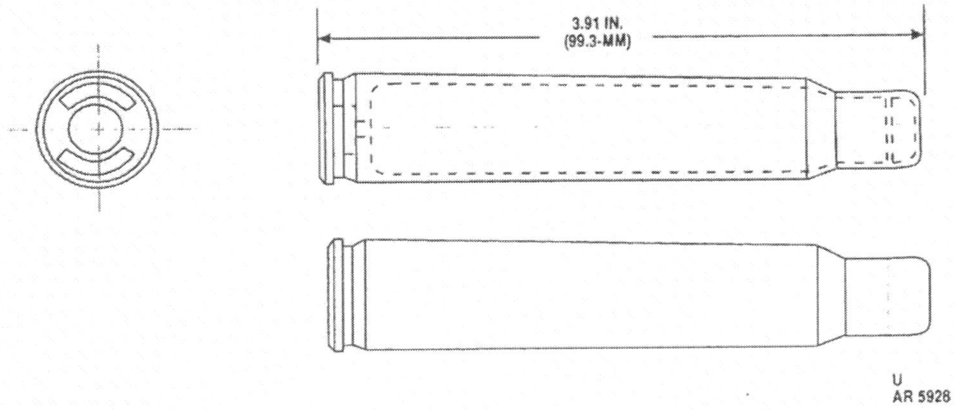


**Type Classification:**

CONT - OTCM 36841.

**Use:**

Machine Gun, Caliber .50, M2 (flexible only). The cartridge is used to simulate firing in training exercises.

**Description:**

BLANK Cartridge. The cartridge is identified by the absence of a bullet and has crimped cartridge case mouth.

**Tabulated Data:**

| | |
|---|---|
| DODAC | 1305-A558 |
| UNO serial number | 0338 |
| UNO proper shipping name | Cartridges for weapons, blank |
| Weight | 917 gr |
| Length | 3.91 in. (99.3 mm) |
| Tracer | NA |
| Primer | Percussion |
| Fuze | NA |

| | |
|---|---|
| Explosive: | |
| Type | NA |
| Weight | NA |
| Incendiary: | |
| Type | NA |
| Weight | NA |
| Propellant: | |
| Type | WC 150 |
| Weight | 46 gr |

**Performance:**

| | |
|---|---|
| Chamber pressure | NA |
| Velocity | NA |

**Shipping and Storage Data:**

| | |
|---|---|
| Quantity-distance class/SCG | 1.4C |
| Storage code | Class V |
| DOT shipping class | C |
| DOT designation | SMALL ARMS AMMUNITION |
| Drawing number | 7673517 |

**References:**

SB 700-20
TM 9-1005-213-10
TM 9-1005-231-10
TM 9-1300-206

THIS PAGE INTENTIONALLY LEFT BLANK

## CARTRIDGE, CALIBER .50, INCENDIARY, M1

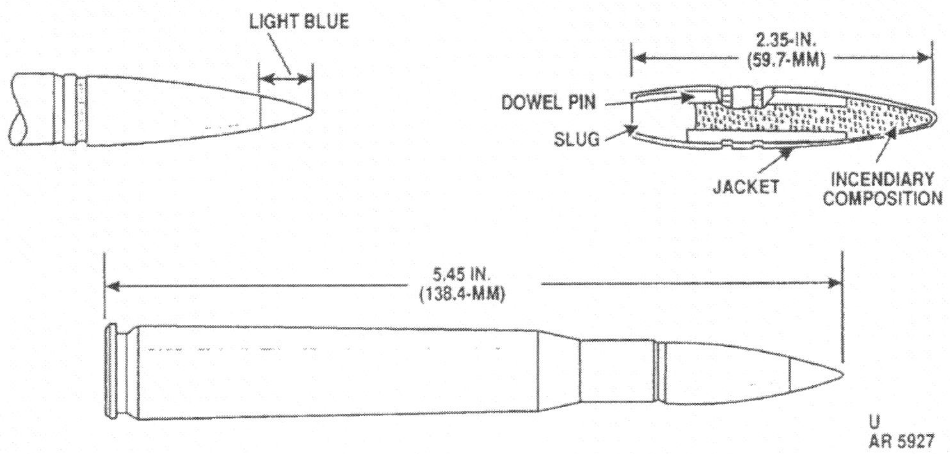


**Type Classification:**

OBS - MSR 11756003.

**Use:**

Machine Guns, Caliber .50, M2 and M85.

**Description:**

INCENDIARY Cartridge. The cartridge is identified by a blue bullet tip.

**Function:**

Upon impact with a hardened or armored target, the incendiary composition bursts into flame and will ignite any flammable material.

**Tabulated Data:**

DODAC ---------------------------------1305-A562
UNO serial number------------------0300
UNO proper shipping
name ----------------------------Ammunition, incendiary
Weight----------------------------------1744 gr
Length ---------------------------------5.45 in, (138.4 mm)
Tracer-----------------------------------NA
Primer ---------------------------------Percussion
Fuze-------------------------------------NA

Explosive:
Type ---------------------------- NA
Weight-------------------------- NA
Incendiary:
Type ---------------------------- IM 11
Weight-------------------------- 34 gr
Propellant:
Type ---------------------------- WC 860
Weight-------------------------- 240 gr

**Performance:**

Chamber pressure------------------ 54,000 psi
Velocity------------------------------- 2950 fps, 78 ft from muzzle

**Shipping and Storage Data:**

Quantity-distance class/
SCG----------------------------- 1.4C
Storage code------------------------ Class V
DOT shipping class----------------- C
DOT designation ------------------- SMALL ARMS AMMUNITION
Drawing number-------------------- 5579812

**References:**

SB 700-20
TM 9-1005-213-10
TM 9-1005-231-10
TM 9-1300-206

THIS PAGE INTENTIONALLY LEFT BLANK

## CARTRIDGE, CAL,IBER .50, TRACER, M1

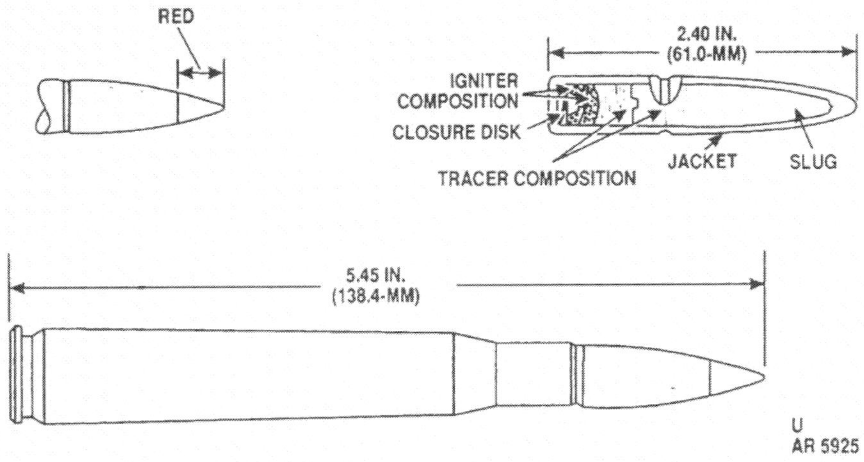


**Type Classification:**

OBS - MSR 11756003.

**Use:**

Machine Guns, Caliber .50, M2 and M85. The tracer is intended to permit visible observation of the bullet's in-flight path or trajectory to the point of impact. Limited to CONUS for training purposes only.

**Description:**

TRACER Cartridge. The cartridge is identified by a red bullet tip.

**Tabulated Data:**

DODAC ------------------------------- 1305-A591
UNO serial number---------------- 0362
UNO proper shipping
name --------------------------- Ammunition, practice
Weight-------------------------------- 1785 gr
Length -------------------------------- 5.45 in. (138.4 mm)
Tracer--------------------------------- R256
Primer -------------------------------- Percussion
Fuze----------------------------------- NA

Explosive:
Type --------------------------- NA
Weight------------------------- NA
Incendiary:
Type --------------------------- NA
Weight------------------------- NA
Propellant:
Type --------------------------- IMR 5010
Weight------------------------- 240 gr

**Performance:**

Chamber pressure---------------- 52,000 psi
Velocity------------------------------ 2700 fps, 78 ft from muzzle

**Shipping and Storage Data:**

Quantity-distance class/
SCG---------------------------- 1.4C
Storage code----------------------- Class V
DOT shipping class--------------- C
DOT designation ------------------ SMALL ARMS AMMUNITION
Drawing number------------------- 5544843

**References:**

SB 700-20
TM 9-1300-206

THIS PAGE INTENTIONALLY LEFT BLANK

**CARTRIDGE, CALIBER .50, BLANK, M1A1**

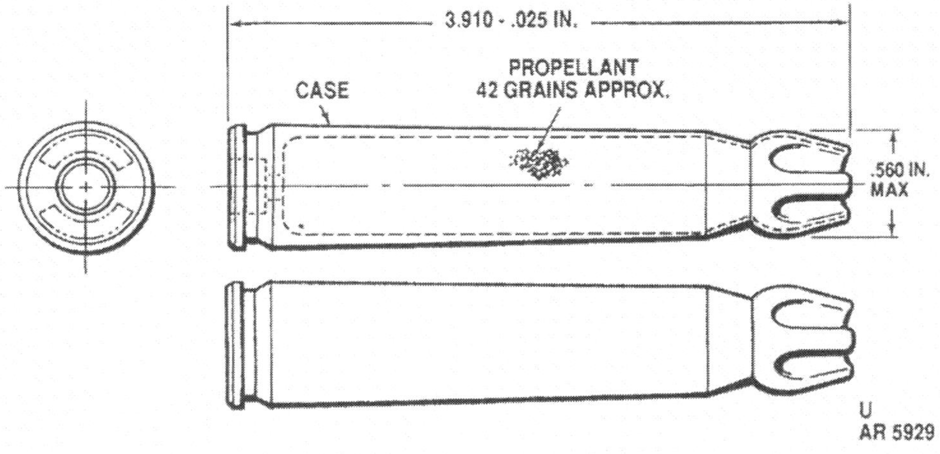


**Type Classification:**

STD - MSR 02806015.

**Use:**

Machine Gun, Caliber .50, M2 with the M19 Blank Ammunition Firing Attachment; and Machine Gun, Caliber .50, M85 with the M20 Blank Firing Attachment. The cartridge is used to simulate firing in training exercises.

**Description:**

BLANK Cartridge. The cartridge is identified by the absence of a bullet. The M1A1 differs from the M1 in that the M1A1 has a rosette crimp at the mouth. Also, the M1A1 is loaded with Dupont Hi Skor 700X propellant.

**Tabulated Data:**

| | |
|---|---|
| DODAC | 1305-A559 |
| UNO serial number | 0338 |
| UNO proper shipping name | Cartridges for weapons, blank |
| Weight | 915 to 955 gr |
| Length | 3.91 in. (99.3 mm) |
| Tracer | NA |
| Primer | Percussion |
| Fuze | NA |
| Explosive: | |
| Type | NA |
| Weight | NA |
| Incendiary: | |
| Type | NA |
| Weight | NA |
| Propellant: | |
| Type | Dupont Hi Skor 700X |
| Weight | 42 gr |

**Performance:**

| | |
|---|---|
| Chamber pressure | NA |
| Velocity | NA |

**Shipping and Storage Data:**

| | |
|---|---|
| Quantity-distance class/ SCG | 1.4S |
| Storage code | Class V |
| DOT shipping class | C |
| DOT designation | SMALL ARMS AMMUNITION |
| Drawing number | 9341551 |

**References:**

DARCOM 700-3-2

THIS PAGE INTENTIONALLY LEFT BLANK

## CARTRIDGE, CALIBER .50, BALL, M2

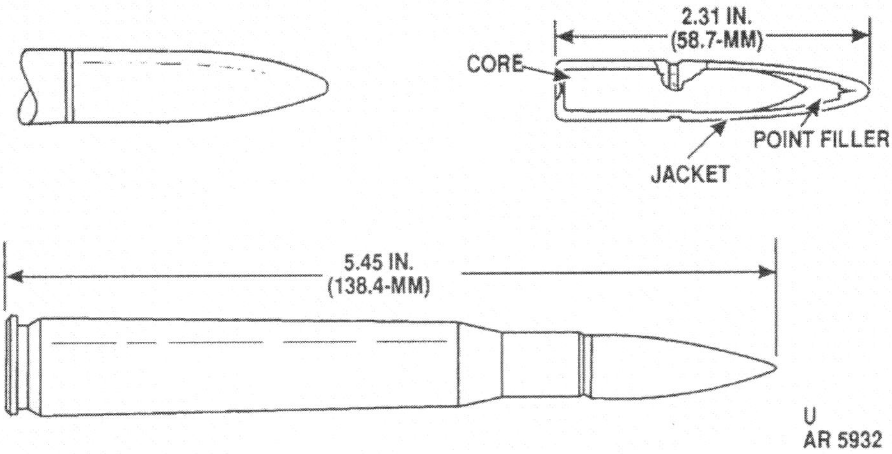


**Type Classification:**

STD - OTCM 36841.

**Use:**

Machine Guns, Caliber .50, M2 and M85. The cartridge is intended for use against personnel or unarmored targets.

**Description:**

BALL Cartridge. The cartridge is identified by a plain bullet.

**Tabulated Data:**

| | |
|---|---|
| DODAC | 1305-A552 |
| UNO serial number | 0339 |
| UNO proper shipping name | Cartridges for weapons, inert projectile |
| Weight | 1813 gr |
| Length | 5.45 in. (138.4 mm) |
| Tracer | NA |
| Primer | Percussion |
| Fuze | |

| | |
|---|---|
| Explosive: | |
| Type | NA |
| Weight | NA |
| Incendiary: | |
| Type | NA |
| Weight | NA |
| Propellant: | |
| Type | WC 860 |
| Weight | 235 gr |

**Performance:**

| | |
|---|---|
| Chamber pressure | 55,000 psi |
| Velocity | 2810 fps, 78 ft from muzzle |

**Shipping and Storage Data:**

| | |
|---|---|
| Quantity-distance class/SCG | 1.4C |
| Storage code | Class V |
| DOT shipping class | C |
| DOT designation | SMALL ARMS AMMUNITION |
| Drawing number | 5577960 |

**References:**

SB 700-20
TM 9-1005-213-10
TM 9-1005-231-10
TM 9-1300-206

THIS PAGE INTENTIONALLY LEFT BLANK

## CARTRIDGE, CALIBER .50, BALL, ARMOR PIERCING, M2

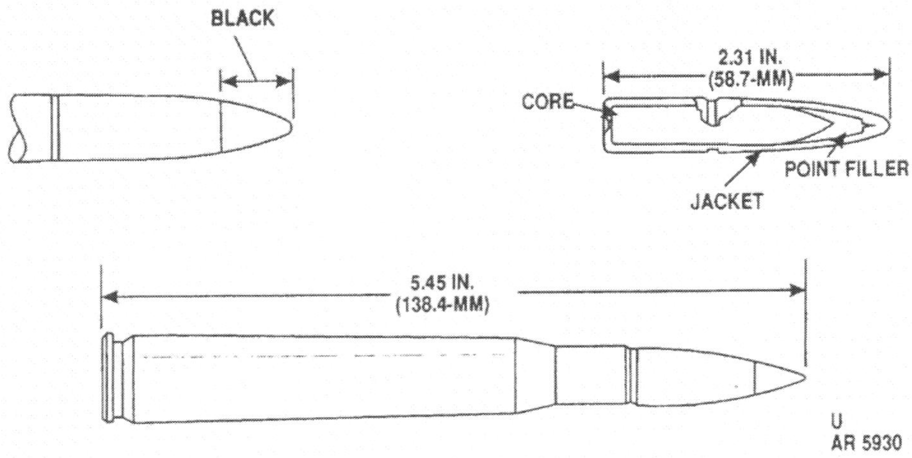


**Type Classification**:

OBS - MSR 11756003.

**Use:**

Machine Guns, Caliber .50, M2 and M85. The cartridge is for use against light-armored or unarmored targets, concrete shelters, and similar bullet-resisting targets.

**Description:**

BALL Cartridge. The cartridge is identified by a black bullet tip.

**Tabulated Data:**

DODAC........................ 1305-A526
UNO serial number ...... 0339
UNO proper shipping
name.......................... Cartridges for weapons, inert projectile
Weight......................... 1812 gr
Length.......................... 5.45 in. (138.4 mm)
Tracer.......................... NA
Primer.......................... Percussion
Fuze ........................... NA

Explosive:
Type ......................... NA
Weight...................... NA
Incendiary:
Type ......................... NA
Weight...................... NA
Propellant:
Type ......................... WC 860
Weight...................... 235 gr

**Performance:**

Chamber pressure........ 53,000 psi
Velocity........................ 2810 fps, 78 ft from muzzle

**Shipping and Storage Data:**

Quantity-distance class/
SCG.......................... 1.4C
Storage code................ Class V
DOT shipping class...... C
DOT designation.......... SMALL ARMS AMMUNITION
Drawing number........... 5569930

**References:**

SB 700-20
TM 9-1005-213-10
TM 9-1005-231-10
TM9-1300-206

THIS PAGE INTENTIONALLY LEFT BLANK

## CARTRIDGE, DUMMY, CALIBER .50, M2

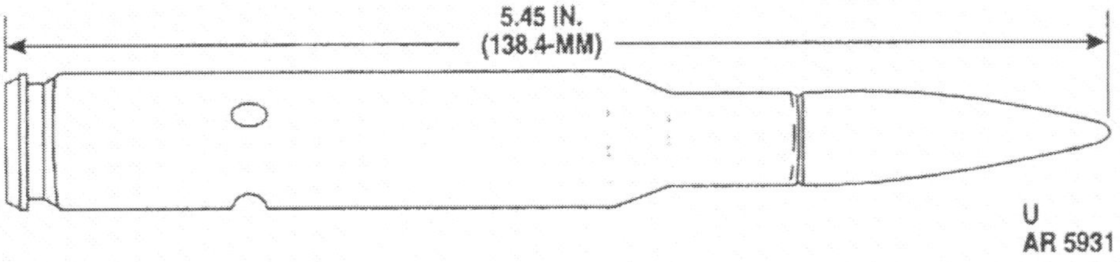


### Type Classification:

STD - OTCM 36841.

### Use:

All caliber .50 weapons except MSC Spotting Gun. The cartridge is used for practice in loading for simulated firing and for inspecting and testing the mechanism of the weapon.

Description:

DUMMY Cartridge. The cartridge is identified by three drilled holes in the cartridge case and the absence of a primer.

### Tabulated Data:

DODAC........................ 1305-A560
Weilt ........................... 1248 gr
Length.......................... 5.45 in. (138.4 mm)
Tracer .......................... NA
Primer.......................... NA
Fuze ........................... NA
Explosive:
  Type ......................... NA
  Weight...................... NA
Incendiary:
  Type ......................... NA
  Weight...................... NA
Propellant:
  Type ......................... NA
  Weight...................... NA

### Performance:

Chamber pressure........ NA
Velocity........................ NA

### Shipping and Storage Data:

Quantity-distance class/
  SCG.......................... NA
Storage code................ NA
DO'I' shipping class...... NA
DOT designation ......... NONEXPLOSIVE AMMUNITION
Drawing number .......... 556579

### References:

SB 700-20
TM 9-1005-213-10
TM 9-1005-231-10
TM 9-1300-206

THIS PAGE INTENTIONALLY LEFT BLANK

## CARTRIDGE, CALIBER .50, BALL, ARMOR PIERCING INCENDIARY, M8

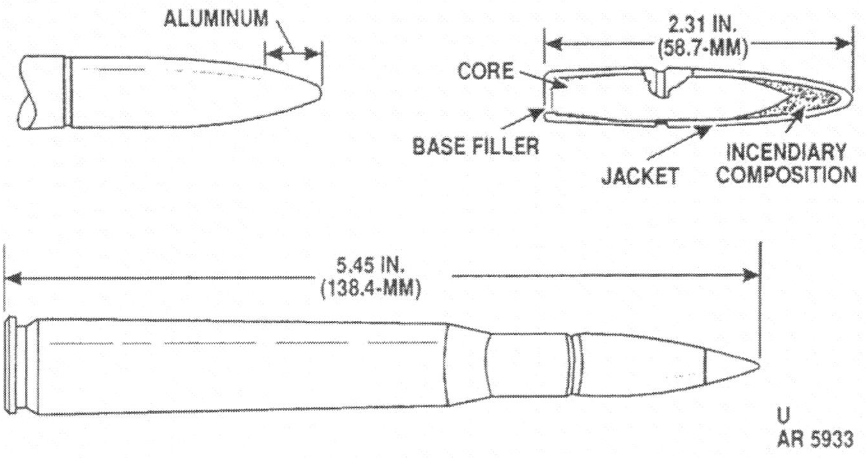


**Type Classification:**

OBS - MSR 11756003.

**Use:**

Machine Guns, Caliber .50, M2 and M85. The cartridge combines the functions of the M2 armor piercing bullet and the incendiary bullet, and is used against flammable targets and light-armored or unarmored targets, concrete shelters, and similar bullet-resisting targets.

**Description:**

BALL Cartridge. The cartridge is identified by an aluminum bullet tip.

**Tabulated Data:**

DODAC........................ 1305-A531
UNO serial number ...... 0300
UNO proper shipping
  name ........................ Ammunition, incendiary
Weight ......................... 1764 gr
Length.......................... 5.45 in. (138.4 mm)
Tracer .......................... NA
Primer.......................... Percussion
Fuze ........................... NA
Explosive:
  Type ......................... NA
  Weight ...................... NA
Incendiary:
  Type ......................... IM 11
Weight ......................... 15 gr
Propellant:
  Type ......................... WC 860
  Weight ...................... 233 gr
Projectile:
  Weight ...................... 622.5 gr

**Performance:**

Chamber pressure........ 59,000 psi
Velocity ........................ 2910 fps, 78 ft from muzzle

**Shipping and Storage Data:**

Quantity-distance class/
  SCG.......................... 1.4G
Storage code................ Class V
DOT shipping class ...... C
DOT designation .......... SMALL ARMS AMMUNITION
Drawing number........... 7970238

**References:**

SB 700-20
TM 9-1005-213-10
TM 9-1005-231-10
TM 9-1300-206

THIS PAGE INTENTIONALLY LEFT BLANK

## CARTRIDGE, CALIBER .50, TRACER, M10

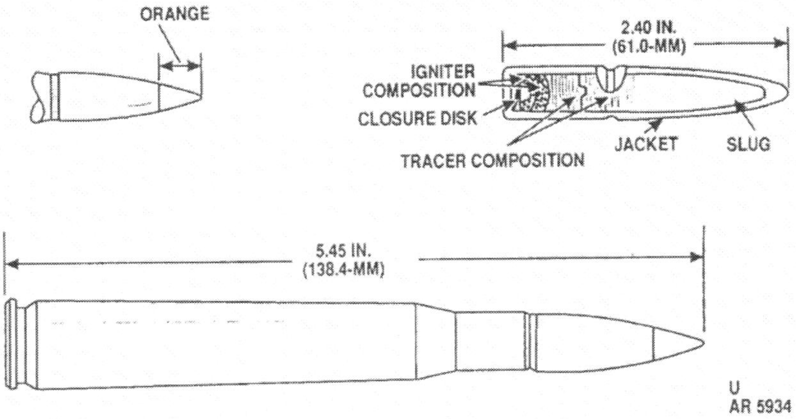


**Type Classification:**

STD - OTCM 37107.

**Use:**

Machine Guns, Caliber .50, M2 and M85.

**Description:**

TRACER Cartridge. The cartridge is identified by an orange bullet tip.

**Function:**

The tracer cartridge exhibits a visible trace from a point not greater than 100 yards from the muzzle of the weapon to a point not less than 1,600 yards from the muzzle.

**Tabulated Data:**

DODAC........................ 1305-A570
UNO serial number ...... 0339
UNO proper shipping
name ........................ Cartridges for weapons, inert projectile
Weight ......................... 1752 gr
Length......... ................ 5.45 in. (138.4 mm)
Tracer .......................... R256
Primer.......................... Percussion

Fuze .......................... NA

Explosive:
Type ......................... NA
Weight ...................... NA
Incendiary:
Type ......................... NA
Weight ...................... NA
Propellant:
Type ......................... IMR 5010
Weight ...................... 240 gr

Performance:

Chamber pressure........ 54,000 psi
Velocity ........................ 2860 fps, 78 ft from muzzle

**Shipping and Storage Data:**

Quantity-distance class/
SCG......................... 1.4C
Storage code................ Class V
DOT shipping class ...... C
DOT designation .......... SMALL ARMS AMMUNITION
Drawing number........... 7670296

**References:**

SB 700-20
TM 9-1005-213-10
TM 9-1005-231-10
TM 9-1300-206

THIS PAGE INTENTIONALLY LEFT BLANK

## CARTRIDGE, CALIBER .50, TRACER, M17

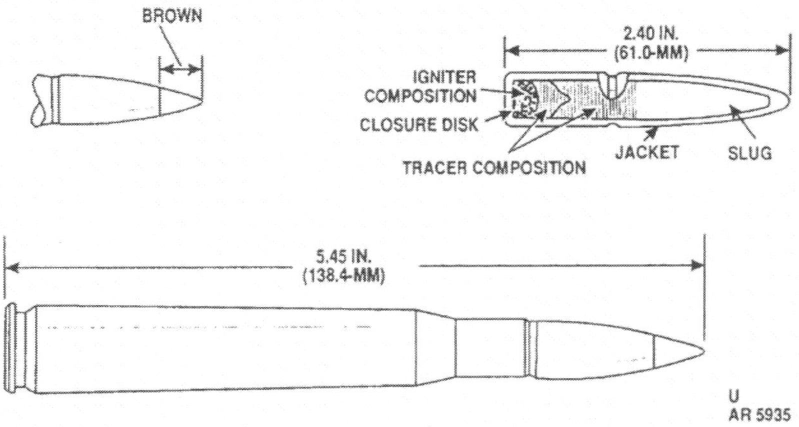


**Type Classification:**

CON - MSR 11756003.

**Use:**

Machine Guns, Caliber .50, M2 and M85. The cartridge tracer is intended to permit visible observation of the bullets in-flight path or trajectory to the point of impact.

**Description:**

TRACER Cartridge. The cartridge is identified by a brown bullet tip.

**Tabulated Data:**

| | |
|---|---|
| DODAC | 1305-A571 |
| UNO serial number | 0339 |
| UNO proper shipping name | Cartridges for weapons, inert projectile |
| Weight | 1732 gr |
| Length | 5.45 in. (138.4 mm) |
| Tracer | R256 |
| Primer | Percussion |
| Fuze | NA |

| | |
|---|---|
| Explosive: | |
| Type | NA |
| Weight | NA |
| Incendiary: | |
| Type | NA |
| Weight | NA |
| Propellant: | |
| Type | IMR 5010 |
| Weight | 225 gr |

**Performance:**

| | |
|---|---|
| Chamber pressure | 54,000 psi |
| Velocity | 2860 fps, 78 ft from muzzle |

**Shipping and Storage Data:**

| | |
|---|---|
| Quantity-distance class/ SCG | 1.4C |
| Storage code | Class V |
| DOT shipping class | C |
| DOT designation | SMALL ARMS AMMUNITION |
| Drawing number | 7672165 |

**References:**

SB 700-20
TM 9-1005-213-10
TM 9-1005-231-10
TM 9-1300-206

THIS PAGE INTENTIONALLY LEFT BLANK

## CARTRIDGE, CALIBER .50, BALL, ARMOR PIERCING INCENDIARY-TRACER, M20

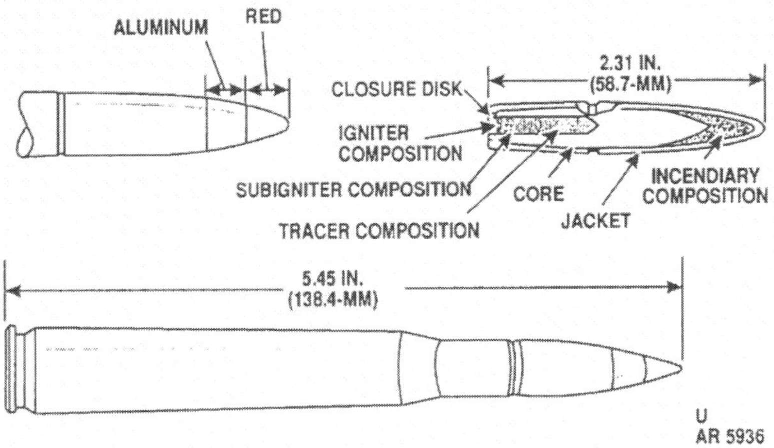


**Type Classification:**

OBS - MSR 04776009.

**Use:**

Machine Guns, Caliber .50, M2 and M85. The cartridge combines the functions of the armor piercing and the incendiary bullet, and is used against flammable targets and light-armored or unarmored targets, concrete shelters, and similar bullet-resisting targets.

**Description:**

BALL Cartridge. The cartridge is identified by a red bullet tip with an aluminum colored ring to the rear of the red tip.

**Tabulated Data:**

DODAC........................ 1305-A542
UNO serial number ...... 0300
UNO proper shipping
name ........................ Ammunition, incendiary
Weight ......................... 1718 gr
Length.......................... 5.45 in. (138.4 mm)
Tracer .......................... R256
Primer.......................... Percussion
Fuze ........................... NA

Explosive:
Type ......................... NA
Weight ...................... NA
Incendiary:
Type ......................... IM 161
Weight ...................... 27 gr
Propellant:
Type ......................... IMR 5010
Weight ...................... 230 gr
Projectile:
Weight ...................... 619 gr

**Performance:**

Chamber pressure........ 55,000 psi
Velocity ........................ 2910 fps, 78 ft from muzzle

**Shipping and Storage Data:**

Quantity-distance class/
SCG.......................... 1.4G
Storage code................ Class V
DOT shipping class ...... C
DOT designation .......... SMALL ARMS AMMUNITION
Drawing number........... 7672003

**References:**

SB 700-20
TM 9-1005-213-10
TM 9-1005-231-10
TM 9-1300-206

THIS PAGE INTENTIONALLY LEFT BLANK

## CARTRIDGE, CALIBER .50, INCENDIARY, M23

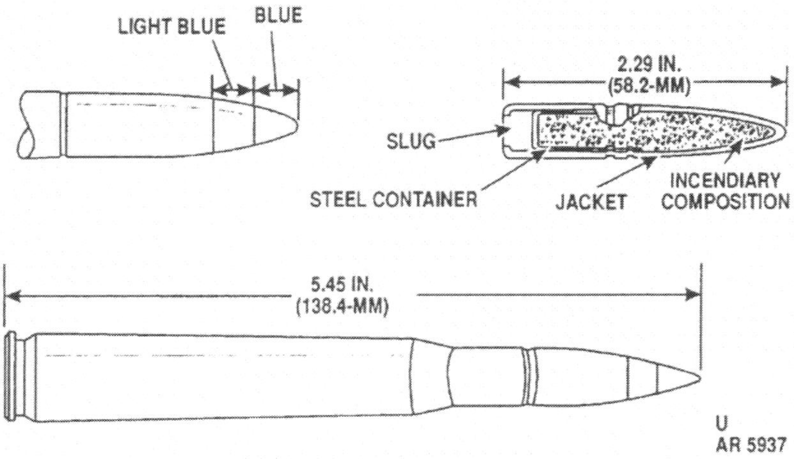


**Type Classification:**

STD - OTCM 36841.

**Use:**

Machine Guns, Caliber .50, M2 and M85. The cartridge is used against flammable targets.

**Description:**

INCENDIARY Cartridge. The cartridge is identified by a blue bullet tip with a light blue ring to the rear of the blue tip.

**Purpose:**

The cartridge is capable of initiating combustion of flammable materials upon target impact at 175 yards.

**Tabulated Data:**

DODAC........................ 1305-A563
UNO serial number ...... 0300
UNO proper shipping
name ........................ Ammunition, incendiary
Weight ......................... 1581 gr
Length.......................... 5.45 in. (138.4 mm)
Tracer .......................... NA

Primer---------------------------- Percussion

Fuze ........................... NA
Explosive:
Type ......................... NA
Weight ...................... NA
Incendiary:
Type ......................... IM 28
Weight ...................... 90 gr
Propellant:
Type ......................... IMR 4831
Weight ...................... 237 gr
Performance:

Chamber pressure........ 58,000 psi
Velocity ........................ 3400 fps, 78 ft from muzzle
Shipping and Storage Data:

Quantity-distance class/
SCG.......................... 1.4G
Storage code................ Class V
DOT shipping class ...... C
DOT designation .......... SMALL ARMS AMMUNITION
Drawing number........... 7673512

**References:**

SB 700-20
TM 9-1005-213-10
TM 9-1005-231-10
TM 9-1300-206

THIS PAGE INTENTIONALLY LEFT BLANK

## CARTRIDGE, CALIBER .50, BALL, M33

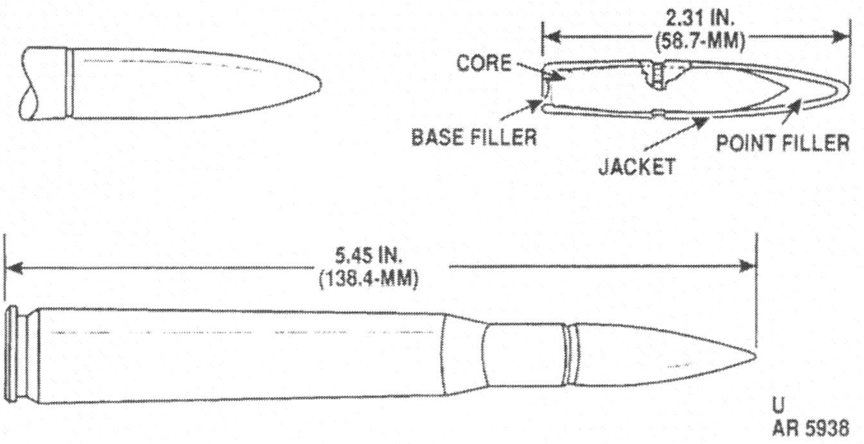


**Type Classification:**

STD - OTCM 36841.

**Use:**

Machine Guns, Caliber .50, M2 and M85. The cartridge is intended for use against personnel or unarmored targets.

**Description:**

BALL Cartridge. The cartridge is identified by a plain bullet tip.

**Tabulated Data:**

| | |
|---|---|
| DODAC | 1305-A552 |
| UNO serial number | 0339 |
| UNO proper shipping name | Cartridges for weapons, inert projectile |
| Weight | 1762 gr |
| Length | 5.45 in. (138.4 mm) |
| Tracer | NA |
| Primer | Percussion |
| Fuze | NA |
| Explosive: | |
| Type | NA |
| Weight | NA |
| Incendiary: | |
| Type | NA |
| Weight | NA |
| Propellant: | |
| Type | WC 860 |
| Weight | 235 gr |

**Performance:**

| | |
|---|---|
| Chamber pressure | 55,000 psi |
| Velocity | 2910 fps, 78 ft from muzzle |

**Shipping and Storage Data:**

| | |
|---|---|
| Quantity-distance class/ SCG | 1.4C |
| Storage code | Class V |
| DOT shipping class | C |
| DOT designation | SMALL ARMS AMMUNITION |
| Drawing number | 7553097 |

**References:**

SB 700-20
TM 9-1005-213-10
TM 9-1005-231-10
TM 9-1300-206

THIS PAGE INTENTIONALLY LEFT BLANK

## CARTRIDGE, CALIBER.50, BALL, SPOTTER-TRACER, M48A1

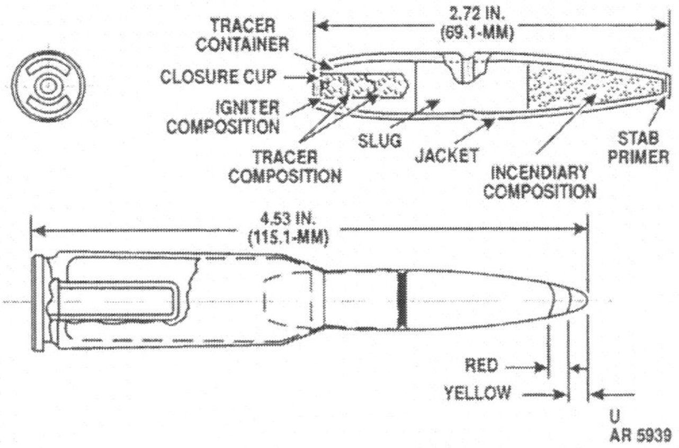


**Type Classification**:

STD - AMCTC 7378.

**Use:**

Rifle, Spotting, Caliber .50, M8C.

**Description:**

BALL Cartridge. The cartridge has specific features which aid in identification. The cartridge case has a distinctive body design which prevents chambering in other weapons. The total length is only 4.5 inches in comparison with 5.5 inches for a standard round. The bullet has a yellow tip with a red ring to the rear of the yellow tip.

**Function:**

In flight, the bullet exhibits a visible trace from a point not greater than 100 yards from the muzzle of the weapon to a point not less than 1,500 yards from the muzzle. The bullet will detonate and produce a flash and a puff of smoke upon impact against a steel plate placed at 175 yards from the muzzle of the weapon at all angles from 90° to 11° with respect to the horizontal.

**Difference Between Models:**

The M48A1 differs from the M48A2 in that the M48A1 has a primer in the point of the bullet.

**Tabulated Data:**

DODAC........................ 1305-A574
UNO serial number ...... 0339
UNO proper shipping
name ........................ Cartridges for weapons, inert projectile
Weight ......................... 1744 gr
Length.......................... 4.53 in. (115.1 mm)
Tracer .......................... R321
Primer.......................... Percussion
Fuze ........................... NA
Explosive:
Type ........................... NA
Weight ......................... NA Incendiary:
Type ......................... IM 144
Weight...................... 32 gr
Propellant:
Type ......................... IMR 7383
Weight...................... 110 gr

**Performance:**

Chamber pressure........ 38,000 psi
Velocity ........................ 1745 fps, 78 ft from muzzle

**Shipping and Storage Data:**

Quantity-distance class/
SCG.......................... 1.4C
Storage code................ Class V

DOT shipping class ...... C
DOT designation .......... SMALL ARMS AMMUNITION
Drawing number........... 8594735

**References:**

SB 700-20
TM 9-1300-206

## CARTRIDGE, CALIBER .50, BALL, SPOTTER-TRACER, M48A2

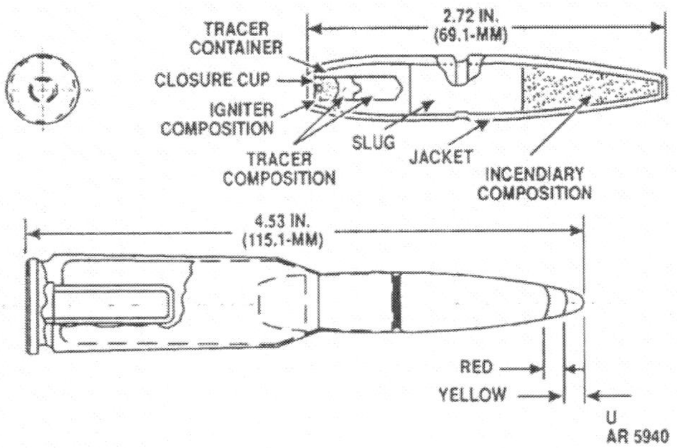


**Type Classification:**

STD - AMCTC 7378.

**Use:**

Rifle, Spotting, Caliber .50, M8C.

**Description:**

BALL Cartridge. The cartridge has specific features which aid in identification. The cartridge case has a distinctive body design which prevents chambering in other weapons. The total length is only 4.5 inches in comparison with 5.5 inches for a standard round. The bullet has a yellow tip with a red ring to the rear of the yellow tip.

**Function:**

In flight, the bullet exhibits a visible trace from a point not greater than 100 yards from the muzzle of the weapon to a point not less than 1,500 yards from the muzzle. The bullet will detonate and produce a flash and a puff of smoke upon impact against a steel plate placed at 175 yards from the muzzle of the weapon at all angles from 90° to 11° with respect to the horizontal.

**Difference Between Models:**

The M48A2 differs from the M48A1 in that the M48A2 does not have a primer in the point of the bullet.

**Tabulated Data:**

DODAC........................ 1305-A574
UNO serial number ..... 0339
UNO proper shipping
name ........................ Cartridges for weapons, inert projectile
Weight ......................... 1744 gr
Length.......................... 4.53 in. (115.1 mm)
Tracer .......................... R321
Primer.......................... Percussion
Fuze ........................... NA
Explosive:
Type ......................... NA
Weight ...................... NA
Incendiary:
Type ......................... IM 144
Weight ...................... 32 gr
Propellant:
Type ......................... IMR 7383
Weight ...................... 110 gr
Projectile:
Weight ...................... 833 gr

**Performance:**

Chamber pressure........ 38,000 psi
Velocity........................ 1745 fps, 78 ft from muzzle

**Shipping and Storage Data:**

Quantity-distance class/
SCG.......................... 1.4C
Storage code................ Class V

DOT shipping class ...... C
DOT designation .......... SMALL ARMS AMMUNITION
Drawing number........... 10542445

**References:**

SB 700-20
TM 9-1300-206

## CARTRIDGE, CALIBER .50, BALL, PLASTIC PRACTICE, M858

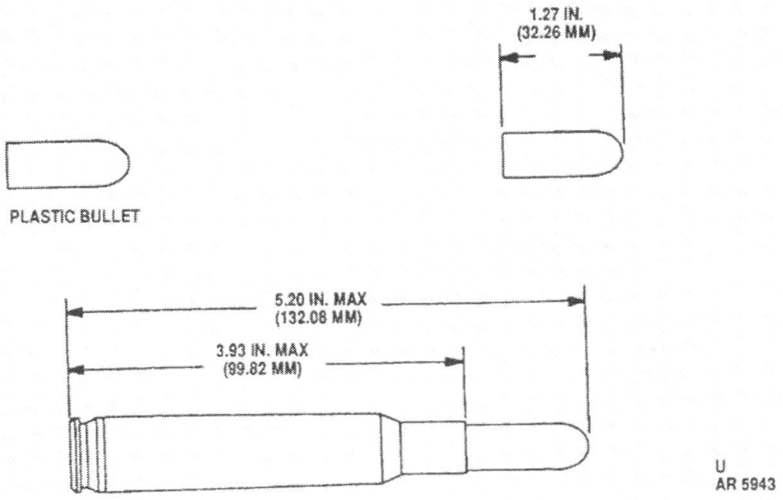


**Type Classification:**

STD - dated 24 May 1983.

**Use:**

Machine Gun, Caliber .50, with the M3 Recoil Amplifier. The cartridge is intended for scaled range training purposes.

**Description:**

BALL Cartridge. The cartridge is identified by the blue bullet and case which are molded into one piece with high density polyethylene plastic.

Tabulated Data:

DODAC........................ 1305-A603
UNO serial number ...... 0012
UNO proper shipping
  name ........................ Cartridges for weapons, inert projectile
Weight ......................... 460 gr
Length.......................... 5.2 in. (132.08 mm)
Tracer .......................... NA
Primer.......................... Percussion
Fuze ........................... NA
Explosive:
  Type ......................... NA
  Weight ...................... NA
Incendiary:
  Type ......................... NA
  Weight ...................... NA
Propellant:
  Type ......................... 10B101
  Weight ...................... 49 gr
Performance:
Chamber pressure........ 26,100 psi
Velocity........................ 2790 fps, 78 ft from muzzle
Maximum range........... 700 m
Shipping and Storage Data:
Quantity-distance class/
  SCG.......................... 1.4S
Storage code................ Class S
DOT shipping class ...... C
DOT designation .......... SMALL ARMS AMMUNITION
Drawing number........... 9340557

**References:**

SB 700-20
TM 9-1005-213-10
TM 9-1005-231-10
TM 9-1300-206

THIS PAGE INTENTIONALLY LEFT BLANK

## CARTRIDGE, CALIBER .50, TRACER, PLASTIC PRACTICE, M860

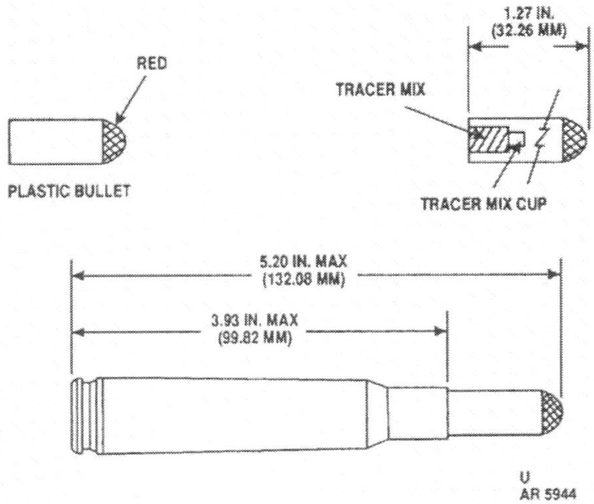


**Type Classification:**

STD - dated 24 May 1983.

**Use:**

Machine Gun, Caliber .50, M2 with the M3 Recoil Amplifier. The cartridge tracer is intended for scaled range training to permit visible observation of the bullet's in-flight path or trajectory to the point of impact. The cartridge is intended for use with the M858 Plastic Practice Ball Cartridge.

**Description:**

TRACER Cartridge. The cartridge is identified by a red bullet tip and the blue bullet and case which are molded into one piece with high density polyethylene plastic.

**Tabulated Data:**

DODAC........................ 1305-A595
UNO serial number ...... 0012
UNO proper shipping name ........................... Cartridges for weapons, inert projectile
Weight ......................... 460 gr
Length.......................... 5.2 in. (132.08 mm)
Tracer .......................... DAG9591254/4
Primer.......................... Percussion
Fuze ........................... NA

Explosive:
Type ......................... NA
Weight ...................... NA
Incendiary:
Type ......................... NA
Weight ...................... NA
Propellant:
Type ......................... 10B101
Weight ...................... 49 gr

**Performance:**

Chamber pressure........ 26,100 psi
Velocity ........................ 2790 fps, 78 ft from muzzle
Maximum range ........... 700 m
Trace ........................... 20-150 m

**Shipping and Storage Data:**

Quantity-distance class/ SCG.......................... 1.4S
Storage code................ Class S
DOT shipping class ...... C
DOT designation .......... SMALL ARMS AMMUNITION
Drawing number........... 9340558

**References:**

SB 700-20
TM 9-1005-213-10
TM 9-1005-231-10
TM 9-1300-206

THIS PAGE INTENTIONALLY LEFT BLANK

## CARTRIDGE, CALIBER .50, SABOTED LIGHT ARMOR PENETRATOR (SLAP), M903

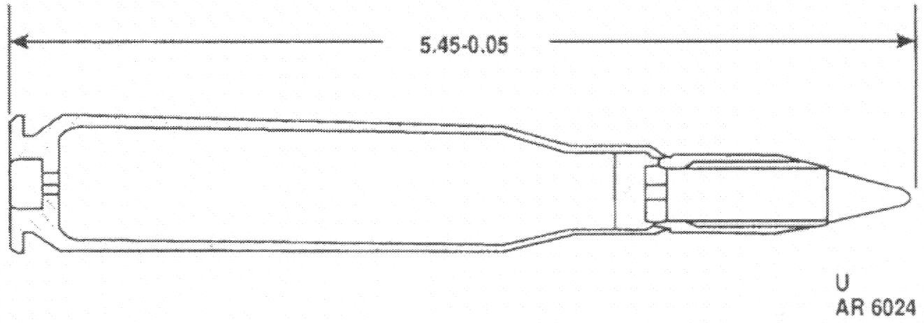


**Type Classification:**

STD - 31 March 1993.

**Use:**

Machine Gun, Caliber .50, M2HB. The cartridge is to be used by soldiers equipped with M2 machine guns.

**Description:**

COMBAT Cartridge. The cartridge is identified by an amber sabot (Ultem 1000).

**Tabulated Data:**

DODAC........................ 1305-A518
UNO serial number ...... 0012
UNO proper shipping
name ........................ Cartridges for weapons, inert projectile
Weight ......................... 1466 gr
Length.......................... 5.45 in. (138.4 m)
Tracer .......................... NA
Primer.......................... Percussion
Fuze ........................... NA
Explosive:
Type ......................... NA
Weight ...................... NA
Incendiary:
Type ......................... NA
Weight ...................... NA
Propellant:
Type ......................... WC 856
Weight ...................... 275 gr
Projectile:
Weight ...................... 355-360 gr

**Performance:**

Chamber pressure........ 55,000 psi
Velocity........................ 4000 fps, 78 ft from muzzle

**Shipping and Storage Data:**

Quantity-distance class/
SCG.......................... 1.4
Storage code................ C
DOT shipping class ...... C
DOT designation .......... CARTRIDGES SMALL ARMS UN0339

THIS PAGE INTENTIONALLY LEFT BLANK

## CARTRIDGE, CALIBER .50, SABOTED LIGHT ARMOR PENETRATOR-TRACER (SLAP-T), M962

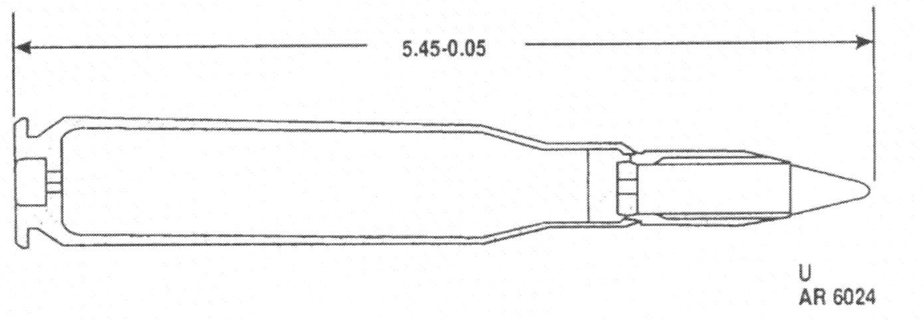


### Type Classification:

STD - 31 March 1993.

### Use:

Machine Gun, Caliber .50, M2HB. The cartridge is to be used by soldiers equipped with M2 machine guns.

### Description:

COMBAT Cartridge. The cartridge is identified by a red sabot (Ultem 1000).

### Tabulated Data:

DODAC........................ 1305-A518
UNO serial number ...... 0012
UNO proper shipping
name ........................ Cartridges for weapons, inert projectile
Weight ......................... 1466 gr
Length.......................... 5.45 in. (138.4 mm)
Tracer .......................... R543
Primer.......................... Percussion
Fuze ........................... NA
Explosive:
Type ......................... NA
Weight ...................... NA
Incendiary:
Type ......................... NA
Weight ...................... NA
Propellant:
Type ......................... WC 856
Weight ...................... 282 gr
Projectile:
Weight ...................... 350-360 gr

### Performance:

Chamber pressure....... 55,000 psi
Velocity ....................... 4000 fps, 78 ft from muzzle

### Shipping and Storage Data:

Quantity-distance class/
SCG.......................... 1.4
Storage code................ C
DOT shipping class ...... C
DOT designation .......... CARTRIDGES SMALL ARMS UN0339

THIS PAGE INTENTIONALLY LEFT BLANK

## CARTRIDGE, CALIBER .50, BALL, PRACTICE, T249E2

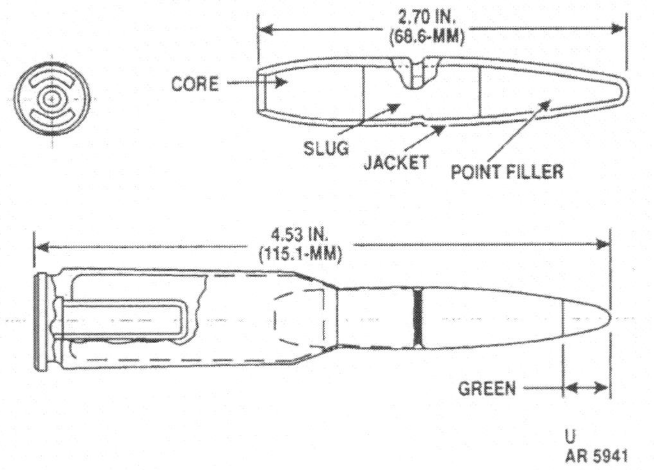


**Type Classification:**

Not applicable.

**Use:**

Rifle, Spotting, Caliber .50, M8C. The cartridge is used as a less expensive substitute for the M48 series spotter-tracer cartridges. The item uses a small amount of Sodium Carbonate Monohydrate as its spotting charge. It is similar to using a flour bag as a bomb.

**Description:**

BALL Cartridge. The cartridge is a duplicate of the M48 series cartridges, a less expensive model used for practice. It is identified by a green bullet tip. It has no tracer or incendiary filler.

**Tabulated Data:**

DODAC........................ 1305-A579
UNO serial number ...... 0339
UNO proper shipping
name ........................ Cartridges for weapons, inert projectile
Weight ......................... 1738 gr
Length.......................... 4.53 in. (115.1 mm)

Tracer .......................... NA
Primer.......................... Percussion

Fuze ........................... NA
Explosive:
Type ......................... NA
Weight ...................... NA
Incendiary:
Type ......................... NA
Weight ...................... NA
Propellant:
Type ......................... IMR 7383
Weight ...................... 110 gr
Projectile:
Weight ...................... 817 gr

**Performance:**

Chamber pressure........ 38,000 psi
Velocity ........................ 1745 fps, 78 ft from muzzle

**Shipping and Storage Data:**

Quantity-distance class/
SCG.......................... 1.4C
Storage code................ Class V
DOT shipping class ...... C
DOT designation .......... SMALL ARMS AMMUNITION
Drawing number........... 10534667

**References:**

SB 700-20
TM 9-1300-206

THIS PAGE INTENTIONALLY LEFT BLANK

## CARTRIDGE, CALIBER .50, BALL, HIGH PRESSURE TEST, T251

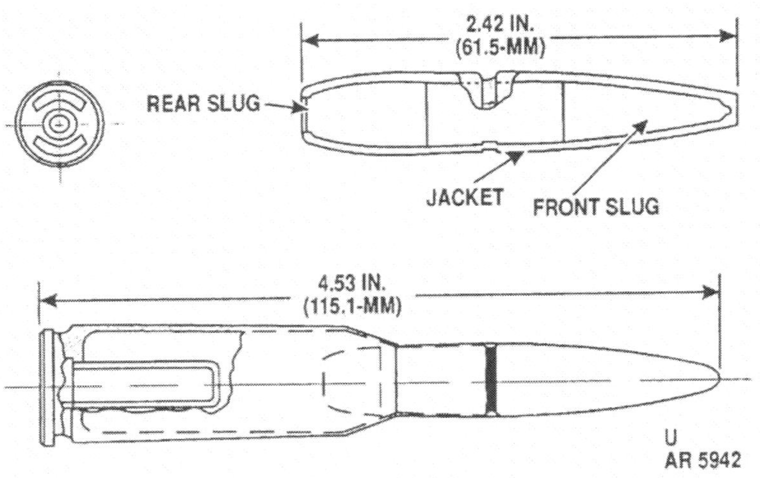


Type Classification:
Not applicable.

**Use:**

Rifle, Spotting, Caliber .50, M8C. The cartridge is used in proof testing the caliber .50 M8C Spotting Rifle during manufacture, test, or repair.

**Description:**

BALL Cartridge. The cartridge is standard small caliber and is identified by a stannic-stained (silvered) cartridge case.

**Tabulated Data:**

DODAC........................ 1305-A578
UNO serial number ...... 0339
UNO proper shipping name............................ Cartridges for weapons, inert projectile
Weight ........................ 1902 gr
Length.......................... 4.53 in. (115.1 mm)
Tracer .......................... NA
Primer.......................... Percussion
Fuze ............................ NA
Explosive:
Type .......................... NA
Weight ....................... NA
Incendiary:
Type ..................................NA
Weight ...............................NA
Propellant:
Type....................................IMR 4831
Weight ............................... 142 gr

**Performance:**

Chamber pressure........ 55,000 psi
Velocity ....................... NA
Shipping and Storage Data:
Quantity-distance class/
SCG ........................... 1.4C
Storage code ............... Class V
DOT shipping class ...... C
DOT designation ......... SMALL ARMS AMMUNITION
Drawing number .......... 10534770

**References:**

SB 700-20
TM 9-1300-206

THIS PAGE INTENTIONALLY LEFT BLANK

## CARTRIDGE, CALIBER .50: ARMOR PIERCING INCENDIARY (API), MK 211 MOD 0

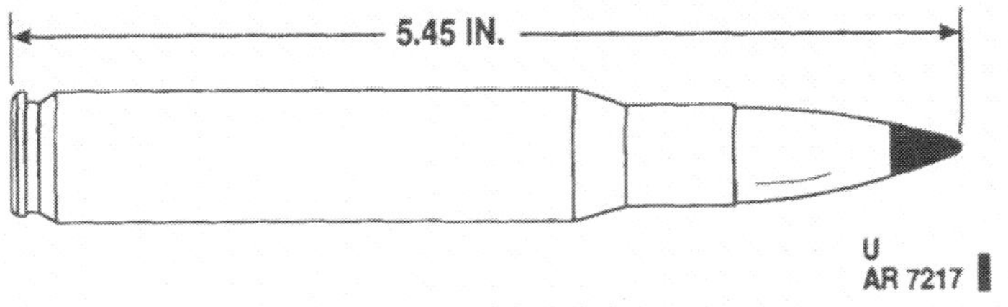


**Type Classification:**

Std - Feb 96 (for U.S. Army).

**Use:**

M2 Machine Gun.

**Description:**

Armor Piercing Cartridge. This cartridge is identified by green bullet tip identification paint. The projectile consists of a brass jacket surrounding a steel body and tungsten core with incendiary and high explosive charges.

**Purpose:**

This cartridge provides improved penetration performance against enemy personnel and light armor vehicles.

**Tabulated Data:**

DODAC (Navy) ................... 1305-01-250-8162, A606; 1305-01-249-6888, A607; 1305-01-251-2581, A608
Weight ...............................1765 gr
Length.................................5.45 in.
Tracer .................................N/A
Primer.................................Percussion
Fuze ...................................N/A

Explosive:
Type ........................ Comp A-4
Weight ...................... 13 gr
Incendiary:
Type ......................... #136
Weight ...................... 13.1 gr
Propellant:
Type ......................... WC 860/MR 5010 or RA-NC-167
Weight ...................... 233 gr
Projectile:
Weight ...................... 671 gr

**Performance:**

Chamber pressure........ 55,000 cup
Velocity ........................ 2910 fps +30 at 78 ft from muzzle

**Shipping and Storage Data:**

DODAC........................ 1305-A607 with M20 API Tracer
Quantity-distance class/SCG..1.4G
Storage code ............... Class C
DOT shipping class ...... C
DOT designation . ........ SMALL ARMS AMMUNITION
Drawing number........... 12576457

**References:**

TM 9-1005-213-10
TM 9-1305-201-20&P
TM 9-1305-201-34&P

THIS PAGE INTENTIONALLY LEFT BLANK

# CHAPTER 10

# 5.56MM CARTRIDGES

THIS PAGE INTENTIONALLY LEFT BLANK

**CARTRIDGE, 5.56MM, BALL, M193**

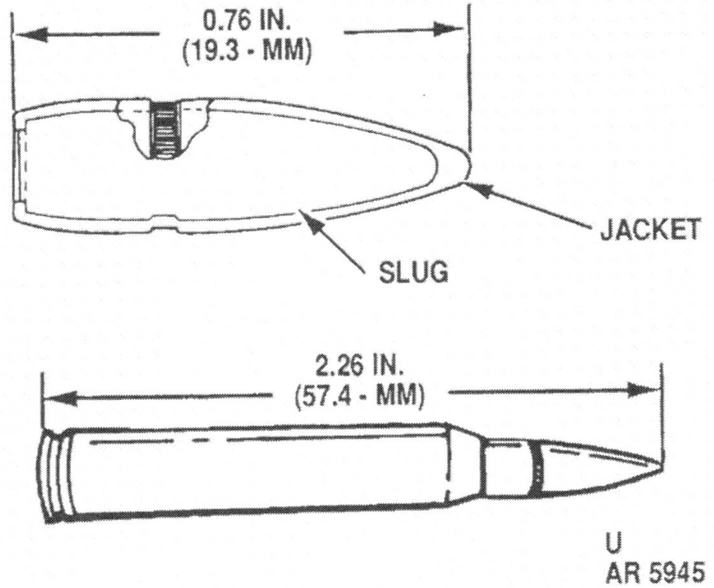


**Type Classification:**

STD - AMCTC 5143.

**Use:**

Rifles, 5.56mm, M16 and M16A1. The cartridge is intended for use against personnel and unarmored targets.

**Description:**

**BALL** Cartridge. The cartridge is identified by a plain bullet tip.

**Tabulated Data:**

DODAC ............................... 1305-A066
UNO serial number ............ 0012
UNO proper shipping
name ............................... Cartridges for weapons, inert projectile
Weight ................................ 182 gr
Length ................................. 2.26 in. (57.4 mm)
Tracer ................................. NA
Primer ................................. Percussion
Fuze .................................... NA
Explosive:
Type ............................... NA
Weight ............................ NA
Incendiary:
Type - .............................. NA
Weight ............................. NA
Propellant:
Type ............................... WC 844 or CMR 170
Weight ............................. 28.5 or 26.5 gr
Projectile:
Weight ............................. 56 gr

**Performance:**

Chamber pressure ............... 52,000 psi
Velocity .............................. 3250 fps, 15 ft from muzzle

**Shipping and Storage Data:**

Quantity-distance class/
SCG ................................ 1.4S
Storage code ...................... Class V

THIS PAGE INTENTIONALLY LEFT BLANK

## CARTRIDGE, 5.56MM, GRENADE, M195

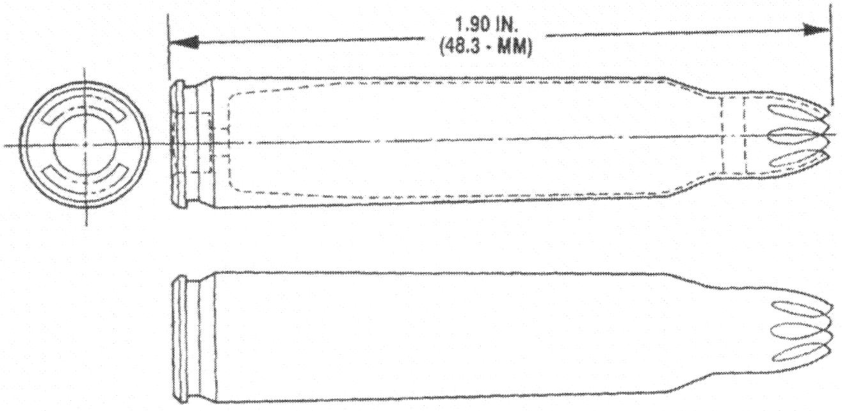


**Type Classification:**

STD - AMCTC 6919.

**Use:**

Rifles, 5.56mm, M16 and M16A1.

**Description:**

GRENADE Cartridge. The cartridge is identified by a rose-petal (rosette-crimp) closure of the cartridge case mouth sealed with red lacquer.

**Function:**

The cartridge provides pressure, on functioning, to project grenades to a desired target using a grenade projection adapter.

**Tabulated Data:**

DODAC ..............................1330-G841
UNO serial number .............0014
UNO proper shipping
name ..............................Cartridges for weapons, blank

Weight ..............................126 gr
Length..............................1.9 in. (48.3 mm)
Tracer ..............................NA
Primer ..............................Percussion
Fuze..................................NA
Explosive:
Type ..............................NA
Weight ............................NA
Incendiary:
Type ..............................NA
Weight ............................NA
Propellant:
Type ..............................IMR 4475
Weight ............................25 gr

**Performance:**

Chamber pressure ..............NA
Velocity ..............................140 to 165 fps, 5 ft 6 in. from muzzle for projected grenade

**Shipping and Storage Data:**

Quantity-distance class/
SCG ..............................1.4S
Storage code ` ....................Class V

THIS PAGE INTENTIONALLY LEFT BLANK

## CARTRIDGE, 5.56MM, TRACER, M196

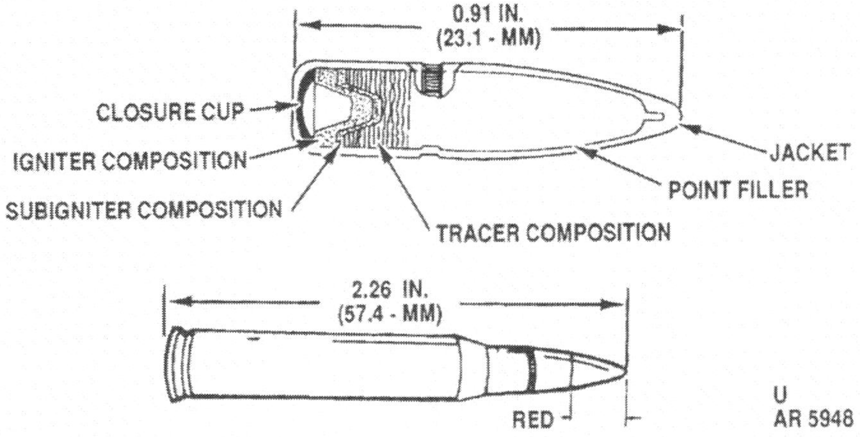


**Type Classification:**

STD - AMCTC 5055.

**Use:**

Rifles, 5.56mm, M16 and M16A1. The tracer is intended to permit visible observation of the bullet's in-flight path or trajectory to the point of impact.

**Description:**

TRACER Cartridge. The cartridge is identified by a red bullet tip.

**Tabulated Data:**

DODAC ............................... 1305-A068
UNO serial number ............ 0012
UNO proper shipping name .................................. Cartridges for weapons, inert projectile
Weight ................................ 177 gr
Length................................. 2.26 in. (57.4 mm)
Tracer ................................. R284
Primer ................................ Percussion
Fuze ................................... NA
Explosive:
Type ............................... NA
Weight ............................ NA
Incendiary:
Type ................................ NA
Weight ............................. NA
Propellant:
Type ............................... WC 844 or IMR 8208M
Weight .............................. 28.5 gr, 25.3 gr or CMR 170, 26.5 gr
Projectile:
Weight ............................ 54 gr

**Performance:**

Chamber pressure .............. 52,000 psi
Velocity .............................. 3200 fps, 15 ft from muzzle

**Shipping and Storage Data:**

Quantity-distance class/ SCG ................................ 1.4S
Storage code ...................... Class V
DOT shipping class ............. C
DOT designation ................ SMALL ARMS AMMUNITION
Drawing number .................. 10534193

**References:**

SB 700-20
TM 9-1005-249-10
TM 9-1300-206

THIS PAGE INTENTIONALLY LEFT BLANK

## CARTRIDGE, 5.56MM, HIGH PRESSURE TEST, M197

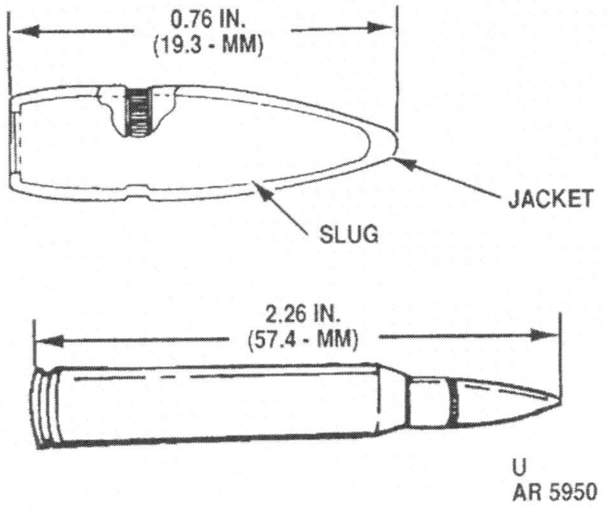


**Type Classification:**

STD - AMCTC 4484.

**Use:**

Rifles, 5.56mm, M16 and M16A1. The cartridge is used to proof test weapons during manufacture, test, or repair.

**Description:**

HIGH PRESSURE TEST Cartridge. The cartridge is identified by a stannic-stained (silvered) or nickel-plated cartridge case.

**Function:**

The cartridge is loaded with a special propellant to produce pressures substantially in excess of the service round.

**Tabulated Data:**

DODAC ................................1305-A070
UNO serial number .............0012
UNO proper shipping name ..................................Cartridges for weapons, inert projectile
Weight ...............................174 gr
Length................................ 2.26 in. (57.4 mm)
Tracer ................................NA
Primer ................................Percussion
Fuze...................................NA
Explosive:
Type ................................NA
Weight ............................NA
Incendiary:
Type ................................NA
Weight ............................NA
Propellant:
Type ...............................SR 7641
Weight- ............................16.7 gr
Projectile:
Weight .............................56 gr

**Performance:**

Chamber pressure...............70,000 psi
Velocity ..............................NA

**Shipping and Storage Data:**

Quantity-distance class/ SCG ................................1.4S
Storage code.......................Class V

THIS PAGE INTENTIONALLY LEFT BLANK

## CARTRIDGE, DUMMY, 5.56MM, M199

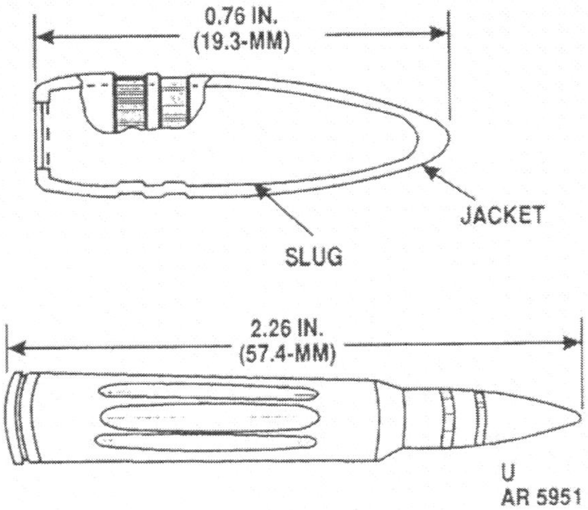


**Type Classification:**

STD - AMCTC 4662.

**Use:**

Rifles, 5.56mm, M16 and M16A1. The cartridge is used for practice in loading 5.56mm weapons, for simulated firing to detect flinching of personnel when firing, and for inspecting and testing the weapon mechanisms.

**Description:**

DUMMY Cartridge. The cartridge is identified by six longitudinal corrugations (flutings) in the cartridge case and there is no primer.

**Tabulated Data:**

DODAC ............ 1305-A060
Weight ............ 150 gr
Length ............ 2.26 in. (57.4 mm)
Tracer ............ NA
Primer ............ NA
Fuze ............ NA
Explosive:
  Type ............ NA
  Weight ............ NA
Incendiary:
  Type ............ NA
  Weight ............ --- NA
Propellant:
  Type ............ NA
  Weight ............ NA

**Performance:**

Chamber pressure ............ NA
Velocity ............ NA

**Shipping and Storage Data:**

Quantity-distance class/
  SCG ............ NA
Storage code ............ NA
DOT shipping class ............ NA
DOT designation ............ NONEXPLOSIVE AMMUNITION
Drawing number ............ 10534021

**References:**

SB 700-20
TM 9-1005-249-10
TM 9-1300-206

THIS PAGE INTENTIONALLY LEFT BLANK

## CARTRIDGE, 5.56MM, BLANK, M200

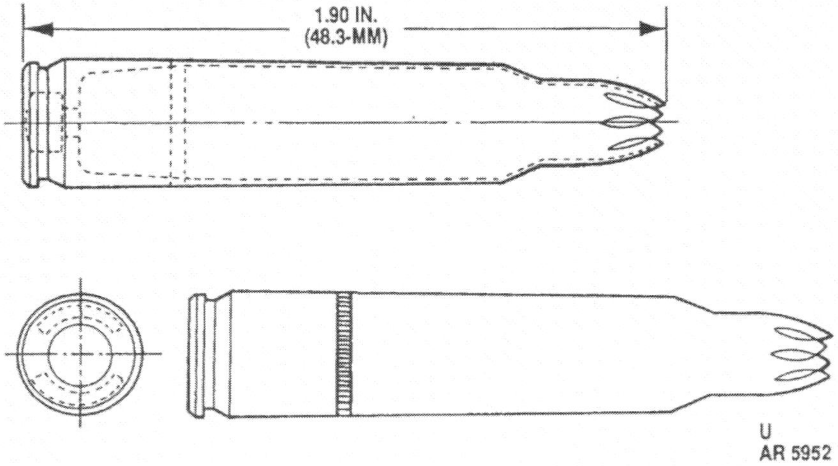


**Type Classification:**

STD - AMCTC 5942.

**Use:**

Rifles, 5.56mm, M16 and M16A1. The cartridge is designed for simulated firing in training exercises and for saluting purposes.

**Description:**

BLANK Cartridge. The cartridge is identified by a rose-petal (rosette-crimp) closure of the cartridge case mouth. An engraved knurl is located 1/2 inch from the head of the cartridge case.

**Tabulated Data:**

DODAC................................1305-A080
UNO serial number .............0014
UNO proper shipping
name ...............................Cartridges for weapons, blank
Weight ...............................107 gr
Length.................................1.9 in. (48.3 mm)
Tracer .................................NA
Primer.................................Percussion
Fuze....................................NA

Explosive:
Type ................................NA
Weight .............................NA
Incendiary:
Type ................................NA
Weight .............................NA
Propellant:
Type ...............................HPC 13
Weight .............................7 gr

**Performance:**

Chamber pressure...............NA
Velocity ..............................NA

**Shipping and Storage Data:**

Quantity-distance class/
SCG ................................1.4S
Storage code ......................Class V
DOT shipping class .............C
DOT designation .................SMALL ARMS AMMUNITION
Drawing number..................10542379

**References:**

SB 700-20
TM 9-1005-249-10
TM 9-1300-206

THIS PAGE INTENTIONALLY LEFT BLANK

## CARTRIDGE, DUMMY, 5.56MM, M232

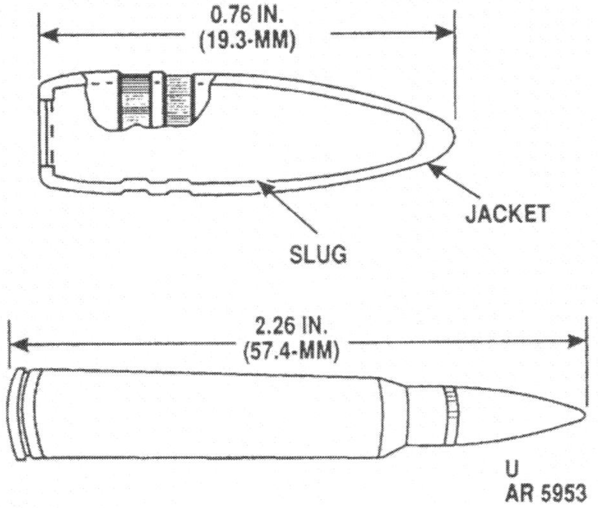


**Type Classification:**

STD - AMCTC 4485.

**Use:**

Rifles, 5.56mm, M16 and M16A1. The cartridge is an inert round used for testing the mechanisms of 5.56 weapons.

**Description:**

DUMMY Cartridge. The cartridge is identified by a black chemical finish over the entire round and by the absence of a primer.

**Tabulated Data:**

DODAC ............1305-A069
Weight ............161 gr
Length ............2.26 in. (57.4 mm)
Tracer ............NA
Primer ............NA
Fuze ............NA
Explosive:
Type ............NA
Weight ............NA
Incendiary:
Type ............NA
Weight ............NA
Propellant:
Type ............NA
Weight ............NA

**Performance:**

Chamber pressure............NA
Velocity ............NA

**Shipping and Storage Data:**

Quantity-distance class/
SCG ............NA
Storage code ............NA
DOT shipping class ............NA
DOT designation ............NONEXPLOSIVE AMMUNITION
Drawing number............10534146

**References:**

SB 700-20
TM 9-1005-249-10
TM 9-1300-206

THIS PAGE INTENTIONALLY LEFT BLANK

## CARTRIDGE, 5.56MMN, BLANK, M755

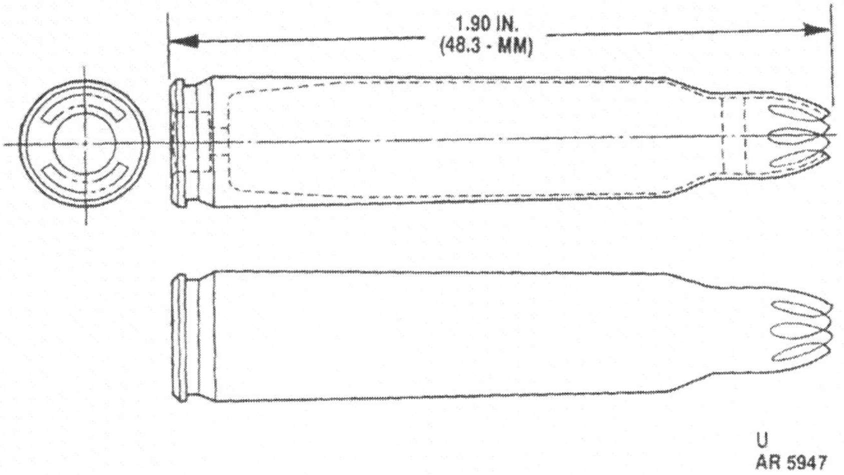


**Type Classification**
STD - MSR 04786005.

**Use:**

The M755 blank cartridge is especially designed for use with the Sting Ring Airfoil Munition System which consists of the M234 launcher and the 64mm riot control M743 projectile fired from the M1GA1 rifle.

**Description:**

BLANK Cartridge. The cartridge is only for use-with - the M234 riot control 64lmm projectile launcher. The launcher is attached to the flash suppressor of the M16A1 rifle and actuated by firing an M755 5.56Gmim blank cartridge. The M755 blank cartridge is the only cartridge that will give the projectile M743 the proper velocity. Use of other blank cartridges may damage the system.

The cartridge is loaded with 12 grains of Dupont Hi Skor 700X propellant closed with a seven petal nose crimp and moisture sealed. The crimped tips of the cartridge is painted with yellow lacquer for identification and thus lessen the chance of mixing them with the standard M16AI rifle blank cartridges.

**Function:**

The blank cartridge is fed into the firing chamber. When fired, the primer ignites the propellant. The expanding propellant ,gases pass fiom the muzzle of the rifle into the launcher manifold. The projectile is ejected from the launcher barrel. The propellant charge is not sufficient to eject the cartridge case which must be ejected by pulling the charging handle of the rifle all the way back.

**Tabulated Data**

DODAC ..............................TBA
Weight ................................112 gr
Length.................................1.9 in. (48.3 mm)
Tracer NA
Primer.................................Percussion
Fuze...................................NA
Explosive:
Type ...............................NA
Weight ............................NA
Incendiary:
Type ...............................NA
Weight ............................NA
Propellant:
Type ...............................Hi Skor 700X
Weight ............................12 gr

**Performance:**

Chamber pressure...............NA
Velocity...............................172 to 198 fps
Maximum range..................100 m

**Shipping and Storage Data:**

Quantity-distance class/
SCG ................................1.4S
Storage code.......................Class I
DOT shipping class.............C
DOT designation ................SMALL ARMS AMMUNITION
Drawing number..................C122-3-43

## CARTRIDGE, 5.56MM, BALL, M855

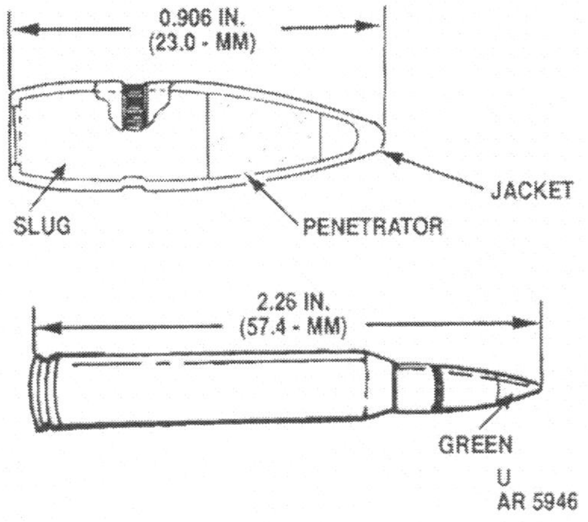


**Type Classification:**

STD - MSR 05826003.

**Use:**

Machine Gun, 5.56mm, M249E1; and Rifle, 5.56mm, M16A2. The cartridge is intended for use against personnel and unarmored targets.

**Description:**

BALL Cartridge. The cartridge is identified by a green bullet tip.

**Tabulated Data:**

DODAC......... .....................1305-A059
UNO serial number .............0012
UNO proper shipping
name..................................Cartridges for weapons, inert projectile
Weight ................................190 gr
Length................................2.26 in. (57.4 mm)
Tracer ................................NA
Primer................................Percussion
Fuze ................................NA
Explosive:
Type ................................NA
Weight ............................NA
Incendiary:
Type ...............................NA
Weight ............................NA
Propellant:
Type ...............................WC 844
Weight ............................26.1 gr
Projectile:
Weight ............................62 gr

**Performance:**

Chamber pressure ..............55,000 psi
Velocity .............................3025 fps, 78 ft from muzzle

**Shipping and Storage Data:**

Quantity-distance class/
SCG................................1.4S
Storage code ......................Class V
DOT shipping class .............C
DOT designation .................SMALL ARMS AMMUNITION
Drawing number .................9342862

**References:**

DARCOM 700-3-2
TM 9-1300-206
TM 9-1305-201-20&P

THIS PAGE INTENTIONALLY LEFT BLANK

## CARTRIDGE, 5.56MM, TRACER, M856

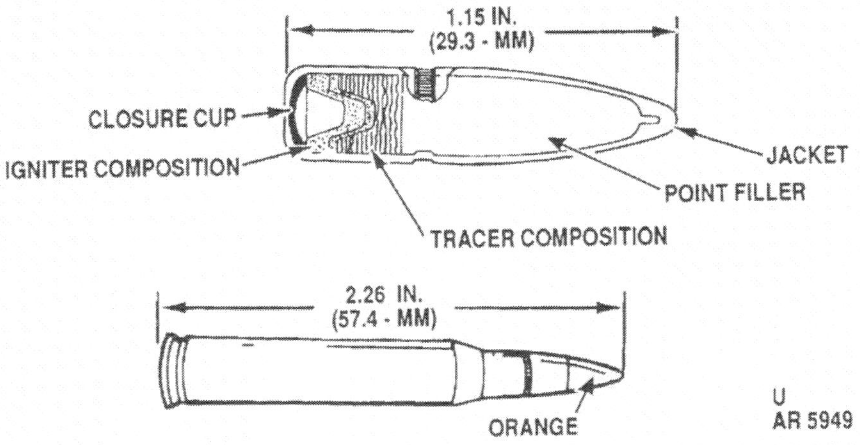


**Type Classification:**

STD - MSR 05826002.

**Use:**

Machine Gun, 5.56mm, M249E1; and Rifle, 5.56mm, M16A2. The tracer is intended to permit visible observation of the bullet's in-flight path or trajectory to the point of impact.

**Description:**

TRACER Cartridge. The cartridge is identified by an orange bullet tip.

**Tabulated Data:**

DODAC............................1305-A063
UNO serial number ............0012
UNO proper shipping name..................................Cartridges for weapons, inert projectile
Weight ...............................191 gr
Length.........................2.26 in. (57.4 mm)
Tracer ..................................NA
Primer..................................Percussion
Fuze ...................................NA

Explosive:
Type ................................NA
Weight ............................NA
Incendiary:
Type ................................NA
Weight ............................NA
Propellant:
Type ................................WC 844
Weight .............................24.7 gr
Projectile:
Weight ............................63.7 gr

**Performance:**

Chamber pressure ..............55,000 psi
Velocity ..............................2870 fps, 78 ft from muzzle

**Shipping and Storage Data:**

Quantity-distance class/
SCG ................................1.4S
Storage code.......................Class V
DOT shipping class .............C
DOT designation .................SMALL ARMS AMMUNITION
Drawing number..................9342863

**References:**

SB 700-20
TM 9-1300-206

THIS PAGE INTENTIONALLY LEFT BLANK

## CARTRIDGE, 5.56MM, PLASTIC, PRACTICE, M862

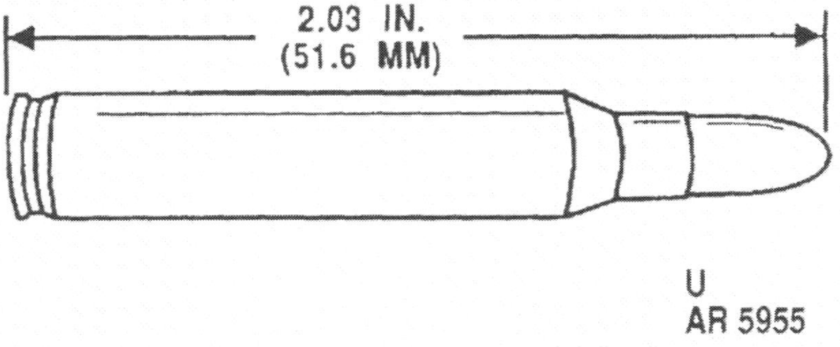


**Type Classification:**

STD - To be approved.

**Use:**

Rifles, 5.56mm, M16A1 and M16A2 with M2 Practice Bolt. The cartridge is for training in local and urban training areas where range restrictions preclude use of full range standard service ammunition.

**Description:**

PLASTIC Cartridge. The cartridge is identified by a blue plastic bullet.

**Tabulated Data:**

| | |
|---|---|
| DODAC | 1305-A065 |
| UNO serial number | 0012 |
| UNO proper shipping name | Cartridges for weapons, inert projectile |
| Weight | 108 gr |
| Length | 2.03 in. (51.6 mm) |
| Tracer | NA |
| Primer | Percussion |
| Fuze | NA |
| Explosive: | |
| Type | NA |
| Weight | NA |
| Incendiary: | |
| Type | NA |
| Weight | NA |
| Propellant: | |
| Type | WPR 260 |
| Weight | 8.7 gr |
| Projectile: | |
| Weight | 3.6 gr |

**Performance:**

| | |
|---|---|
| Chamber pressure | 15,750 psi |
| Velocity | 4525 fps, 15 ft from muzzle |

**Shipping and Storage Data:**

| | |
|---|---|
| Quantity-distance class/ SCG | 1.4S |
| Storage code | Class V |
| DOT shipping class | C |
| DOT designation | SMALL ARMS AMMUNITION |
| Drawing number | 12598589 |

**References:**

SB 700-20
TM 9-1005-249-10
TM 9-1300-206

THIS PAGE INTENTIONALLY LEFT BLANK

## CARTRIDGE, 5.56MM, ARMOR PIERCING (AP), M995

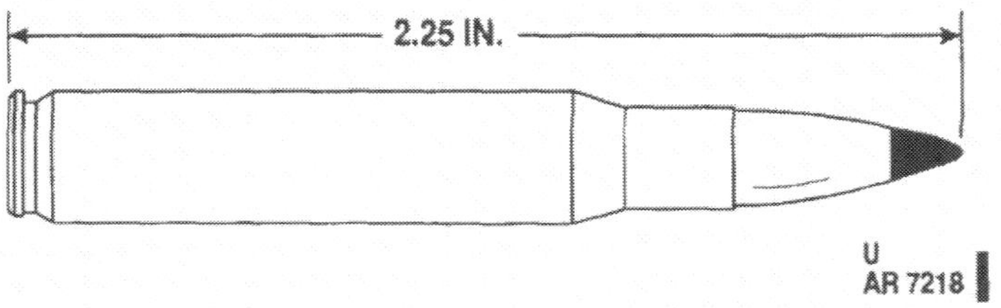


**Type Classification:**

Std - 29 Mar 96.

**Use:**

Rifle, 5.56MM, M16A2 Rifle, 5.56MM, M249 Squad Automatic Weapon.

**Description:**

Armor Piercing Cartridge. This cartridge is identified by black bullet tip identification paint.

**Purpose:**

This cartridge provides improved penetration performance against various targets as compared to standard ball ammunition.

**Tabulated Data:**

DODAC..... .................. 1305-AA01, 1305-AA02
Weight ......................... 180 gr
Length.......................... 2.25 in.
Tracer .......................... N/A
Primer.......................... 4.5mm Berdan
Fuze............................. N/A
Explosive:
  Type ......................... N/A
  Weight ....................... N/A
Incendiary:
  Type ..................................... N/A
  Weight .................................. N/A
Propellant:
  Type ..................................... WCR845
  Weight .................................. 27.5 gr
Projectile:
  Weight .................................. 52 gr

**Performance:**

Chamber pressure..................... 50,250 psi
Velocity .................................... 3324 fps at 78 ft from muzzle

Shipping and Storage Data:

Linked AP cartridge................... 1305-01-393-7050, AA01
Linked AP cartridge with tracer.. 1305-01-393-7052, AA02
Quantity-distance class/SCG..... 1.4S
Storage code............................. Class C
DOT shipping class ................... C
DOT designation ....................... SMALL ARMS AMMUNITION
Drawing number........................ 12956131

**References:**

TM 9-1005-201-10
TM 9-1005-319-10
TM 9-1305-201-20&P
TM 9-1305-201-34&P

THIS PAGE INTENTIONALLY LEFT BLANK

# CHAPTER 11

# 7.62MM CARTRIDGES

THIS PAGE INTENTIONALLY LEFT BLANK

## CARTRIDGE, 7.62MM, BALL, M59

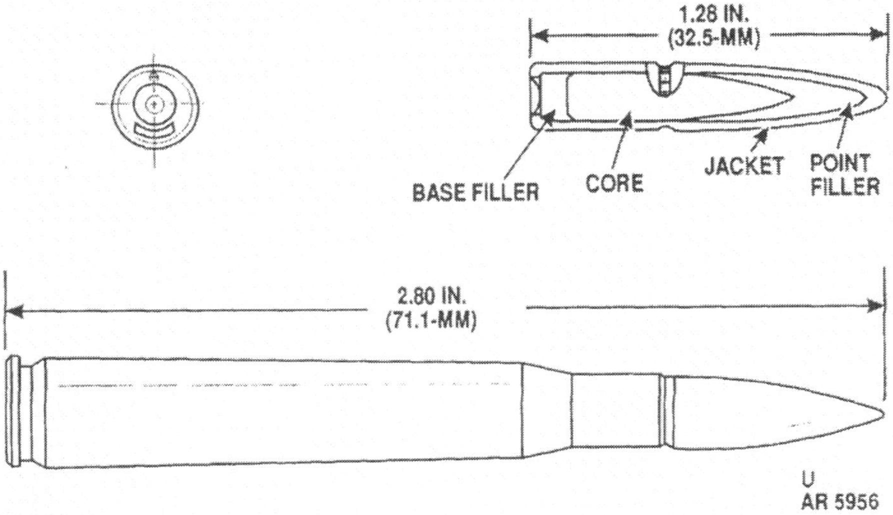


**Type Classification:**

STD - OTCM 36841.

**Use:**

Machine Guns, 7.62mm, M60 and M219; and Rifle, 7.62mm, M14. The cartridge is intended for use against personnel and unarmored targets.

**Description:**

BALL Cartridge. The cartridge is identified by a plain bullet tip.

**Tabulated Data:**

DODAC ........................ 1305-A143
UNO serial number ..... 0012
UNO proper shipping
name ........................ Cartridges for weapons, inert projectile
Weight ........................ 393 gr
Length.......................... 2.8 in. (71.1 mm)
Tracer ............ ............. NA
Primer ......................... NA

Fuze ............................ NA
Explosive:
Type ......................... NA
Weight ...................... NA
Incendiary:
Type ......................... NA
Weight ...................... NA
Propellant:
Type ........................ WC 846
Weight ..................... 46 gr
Projectile:
Weight ...................... 150.5 gr

**Performance:**

Chamber pressure ....... 50,000 psi
Velocity ....................... 2750 fps, 78 ft from muzzle

**Shipping and Storage Data:**

Quantity-distance class/
SCG ......................... 1.4S
Storage code ............... Class V
DOT shipping class ..... C
DOT designation ......... SMALL ARMS AMMUNITION
Drawing number .......... 7553702

**References:**

SB 700-20
TM 9-1005-223-10
TM 9-1005-223-12P
TM 9-1005-224-10
TM 9-1005-233-10
TM 9-1005-298-12
TM 9-1005-313-10
TM 9-1300-206

## CARTRIDGE, 7.62MM, HIGH PRESSURE TEST, M60

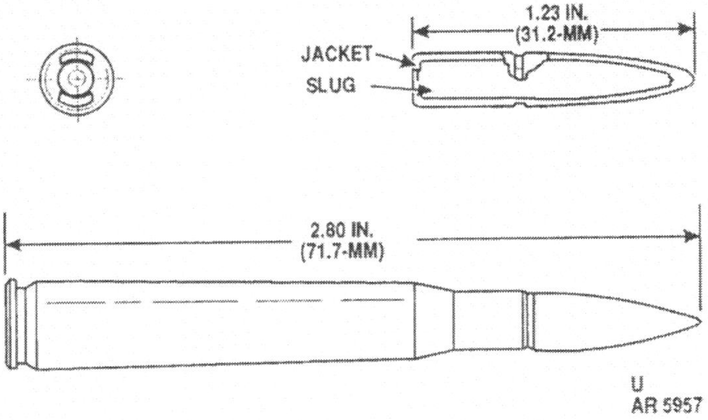


**Type Classification:**

OBS - MSR 11756003.

**Use:**

All 7.62mm weapons. The cartridge is not for field issue, but is used for proof firing of weapons during manufacture, test, or repair.

**Description:**

HIGH PRESSURE TEST Cartridge. The cartridge is identified by a stannic-stained (silvered) case.

**Function:**

The cartridge is loaded with a special propellant to produce pressures substantially in excess of the service cartridge.

**Tabulated Data:**

DODAC........................ 1305-A129
UNO serial number ..... 0012
UNO proper shipping
name ........................ Cartridges for weapons, inert projectile
Weight ........................ 412 gr
Length ........................ 2.8 in. (71.1 mm)
Tracer ........................ NA
Primer ........................ Percussion
Fuze ........................ NA
Explosive:
Type ........................ NA
Weight ........................ NA
Incendiary:
Type ........................ NA
Weight ........................ NA
Propellant:
Type ........................ IMR 4475
Weight ........................ 41 gr
Projectile:
Weight ........................ 171.5 gr

**Performance:**

Chamber pressure ....... 67,500 psi
Velocity ........................ NA

**Shipping and Storage Data:**

Quantity-distance class/
SCG ........................ 1.4S
Storage code ............... Class V
DOT shipping class ..... C
DOT designation ......... SMALL ARMS AMMUNITION
Drawing number .......... 7553703

**References:**

SB 700-20
TM 9-1005-223-10
TM 9-1005-223-12P
TM 9-1005-224-10
TM 9-1005-233-10
TM 9-1005-298-12
TM 9-1005-313-10
TM 9-1300-206

THIS PAGE INTENTIONALLY LEFT BLANK

## CARTRIDGE, 7.62MM, ARMOR PIERCING, M61

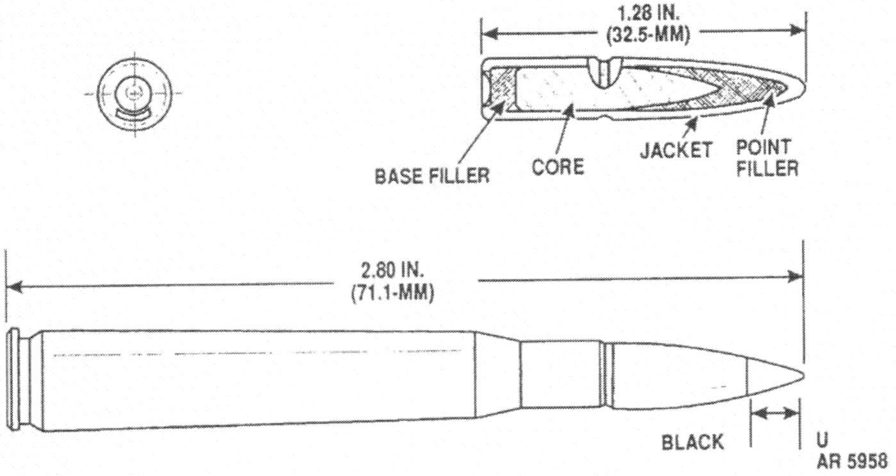


**Type Classification:**

OBS - MSR 11756003.

**Use:**

Machine Guns, 7.62mm, M60, M219 and M240; and Rifle, 7.62mm, M14. The cartridge is used in rifles and machine guns against personnel and light armored or unarmored targets, concrete shelters, and similar bullet-resisting targets.

**Description:**

ARMOR PIERCING Cartridge. The cartridge is identified by a black bullet tip.

**Tabulated Data:**

DODAC........................ 1305-A120
UNO serial number ..... 0012
UNO proper shipping
  name ....................... Cartridges for weapons, inert projectile
Weight ........................ 393 gr
Length ......................... 2.8 in. (71.1 mm)
Tracer .......................... NA
Primer ......................... Percussion
Fuze ............................ NA
Explosive:
  Type ......................... NA
  Weight ...................... NA
Incendiary:
  Type ......................... NA
  Weight ...................... NA
Propellant:
  Type ......................... IMR 4475
  Weight ...................... 41 gr
Projectile:
  Weight ...................... 150.5 gr

**Performance:**

Chamber pressure ....... 50,000 psi
Velocity ....................... 2750 fps, 78 ft from muzzle

**Shipping and Storage Data:**

Quantity-distance class/
  SCG ......................... 1.4S
Storage code ............... Class V
DOT shipping class ...... C
DOT designation ......... SMALL ARMS AMMUNITION
Drawing number .......... 7553704

**References:**

SB 700-20
TM 9-1005-223-10
TM 9-1005-223-12P
TM 9-1005-224-10
TM 9-1005-233-10
TM 9-1005-298-12
TM 9-1005-313-10
TM 9-1300-206

## CARTRIDGE, 7.62MM, TRACER, M62

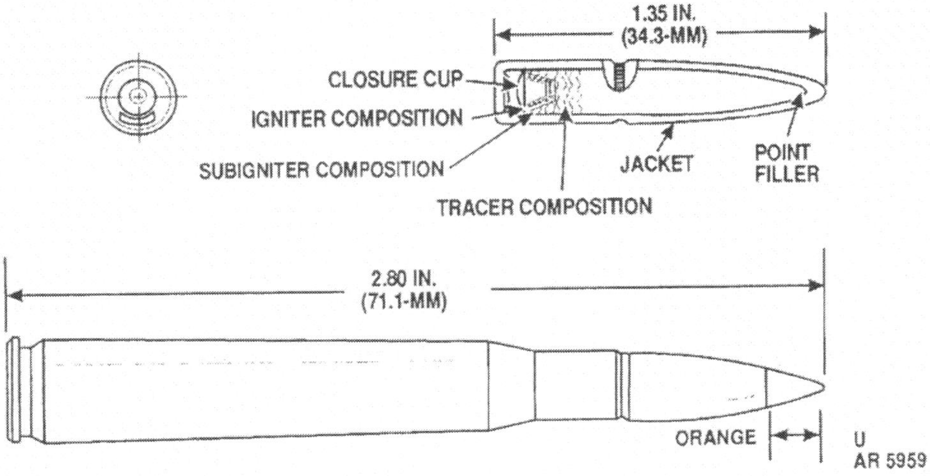


### Type Classification:

CON - MSR 11756003.

### Use:

Machine Guns, 7.62mm, M60, M219 and M240; and Rifle, 7.62mm, M14. The tracer is intended to permit visible observation of the bullet's in-flight path or trajectory to the point of impact.

### Description:

TRACER Cartridge. The cartridge is identified by an orange bullet tip.

### Tabulated Data:

DODAC......... .............. 1305-A124
UNO serial number ..... 0012
UNO proper shipping
name ....................... Cartridges for weapons, inert projectile
Weight ........................ 383 gr
Length.......................... 2.8 in. (71.1 mm)
Tracer .......................... R284

Primer.......................... Percussion
Fuze ............................ NA
Explosive:
Type ........................ NA
Weight ..................... NA
Incendiary:
Type ........................ NA
Weight...................... NA
Propellant:
Type ........................ WC 846
Weight ..................... 46 gr
Projectile:
Weight ..................... 142 gr

### Performance:

Chamber pressure ....... 50,000 psi
Velocity ....................... 2750 fps, 78 ft from muzzle

### Shipping and Storage Data:

Quantity-distance class/
SCG ......................... 1.4S
Storage code ............... Class V
DOT shipping class ..... C
DOT designation ......... SMALL ARMS AMMUNITION
Drawing number .......... 10522000

**References:**

SB 700-20
TM 9-1005-223-10
TM 9-1005-223-12P
TM 9-1005-224-10
TM 9-1005-233-10
TM 9-1005-298-12
TM 9-1005-313-10
TM 9-1300-206

CARTRIDGE, 7.62MM, TRACER, M62 (OVERHEAD FIRE MISSION)

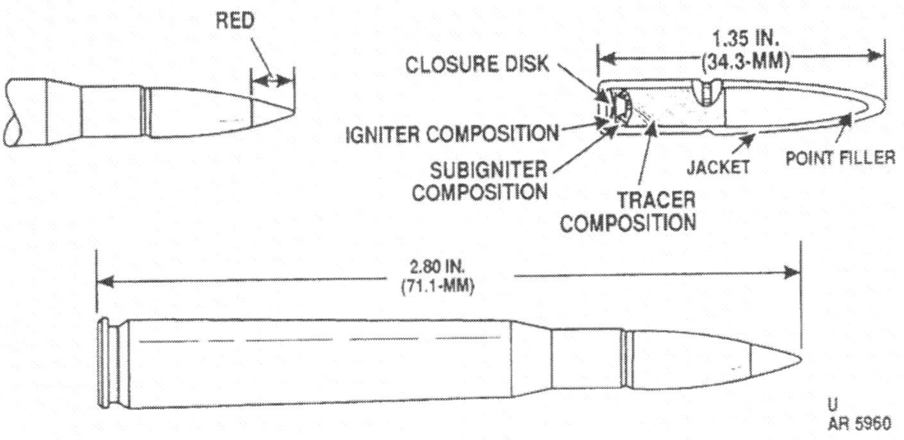


**Type Classification:**

CON - MSR 11756003.

**Use:**

Machine Guns, 7.62mm, M60, M219 and M240; and Rifle, 7.62mm, M14. The cartridge is used in weapons for firing over the heads of troops being trained in field exercises. Stringent production control and screening of ammunition lots ensure the safety of personnel operating immediately below the trajectory of the fired bullets.

**Description:**

TRACER Cartridge. The cartridge is identified by a red bullet tip.

**Tabulated Data:**

DODAC ........................ 1305-A167
UNO serial number ..... 0012
UNO proper shipping
name ........................ Cartridges for weapons, inert projectile
Weight ........................ 387 gr
Length ........................ 2.8 in. (71.1 mm)
Tracer ........................ R284
Primer ........................ Percussion
Fuze ........................ NA
Explosive:
Type ........................ NA
Weight ........................ NA
Incendiary:
Type ........................ NA
Weight ........................ NA
Propellant:
Type ........................ WC 846
Weight ........................ 46 gr
Projectile:
Weight ........................ 146 gr

**Performance:**

Chamber pressure ....... 50,000 psi
Velocity ........................ 2750 fps, 78 ft from muzzle

**Shipping and Storage Data:**

Quantity-distance class/
SCG ........................ 1.4S
Storage code ............... Class V
DOT shipping class ...... C
DOT designation ......... SMALL ARMS AMMUNITION
Drawing number .......... 10535493

**References:**

SB 700-20
TM 9-1005-223-10
TM 9-1005-223-12P
TM 9-1005-224-10
TM 9-1005-233-10
TM 9-1005-298-12
TM 9-1005-313-10
TM 9-1300-206

THIS PAGE INTENTIONALLY LEFT BLANK

## CARTRIDGE, DUMMY, 7.62MM, M63

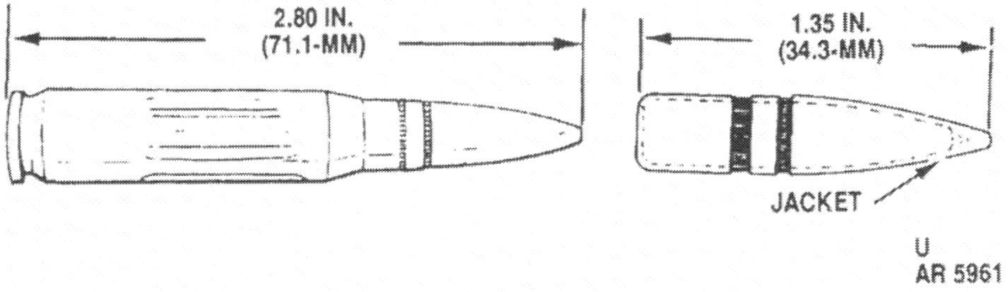


**Type Classification:**

STD - OTCM 36841.

**Use:**

Machine Guns, 7.62mm, M60, M219 and M240; and Rifle, 7.62mm, M14. The cartridge is used for practice in loading 7.62mm weapons for simulated firing to detect flinching of personnel during firing and for inspecting and testing the weapon mechanism.

**Description:**

DUMMY Cartridge. The cartridge is identified by six longitudinal corrugations (flutings) on the cartridge case. There is no primer and no vent hole in the primer pocket.

**Tabulated Data:**

DODAC .................... 1305-A135
Weight ........................ 258 gr
Length ......................... 2.8 in. (71.1 mm)
Tracer .......................... NA
Primer ......................... NA
Fuze ............................ NA
Explosive:
  Type ........................ NA
  Weight...................... NA
Incendiary:
  Type ........................ NA
  Weight ..................... NA
Propellant:
  Type ........................ NA
  Weight ..................... NA
Projectile:
  Weight...................... NA

**Performance:**

Chamber pressure ....... NA
Velocity ....................... NA

**Shipping and Storage Data:**

Quantity-distance class/
  SCG ......................... NA
Storage code ............... NA
DOT shipping class ..... NA
DOT designation ......... NONEXPLOSIVE AMMUNITION
Drawing number .......... 7553706

**References:**

SB 700-20
TM 9-1005-223-10
TM 9-1005-223-12P
TM 9-1005-224-10
TM 9-1005-233-10
TM 9-1005-298-12
TM 9-1005-313-10
TM 9-1300-206

THIS PAGE INTENTIONALLY LEFT BLANK

## CARTRIDGE, 7.62MM, GRENADE, M64

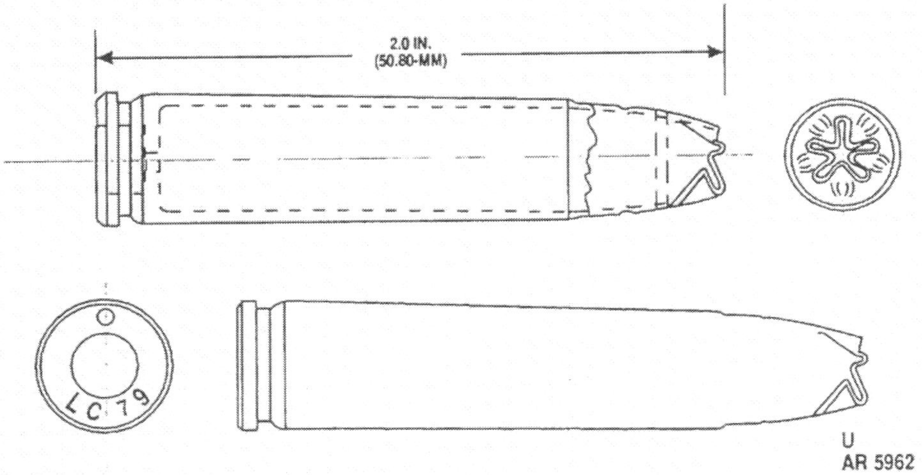


**Type Classification:**

OBS - MSR 11756003.

**Use:**

Rifle, 7.62mm, M14.

**Description:**

GRENADE Cartridge. The cartridge is identified by a rose-petal (rosette-crimp) closure of the cartridge case mouth and sealed with red lacquer.

**Function:**

The cartridge provides pressure upon functioning to project rifle grenade to a desired target when using a grenade projectile adapter and dragon missile launch effect trainer (LET).

**Tabulated Data:**

DODAC........................ 1330-G839
UNO serial number ..... 0014
UNO proper shipping
name ....................... Cartridges for weapons, blank
Weight ........................ 295 gr
Length ......................... 2 in. (50.8 mam)
Tracer .......................... NA
Primer ......................... Percussion
Fuze ............................ NA
Explosive:
Type ........................ NA
Weight ..................... NA
Incendiary:
Type ........................ NA
Weight ..................... NA
Propellant:
Type ........................ WC 8:30
Weight ...................... 45 gr
Projectile:
Weight ...................... NA

**Performance:**

Chamber pressure ....... NA
Velocity ....................... 160 fps, 5.(, ft from nuzzle for projected grenade

**Shipping and Storage Data:**

Quantity-distance class/
SCG ........................ 1.4S
Storage code ............... Class V

DOT shipping class...... C
DOT designation ......... SMALL ARMS AMMUNITION
Drawing number .......... 7553707

**References:**

SB 700-20
TM 9-1300-206

CARTRIDGE, 7.62MM, BAL., M80

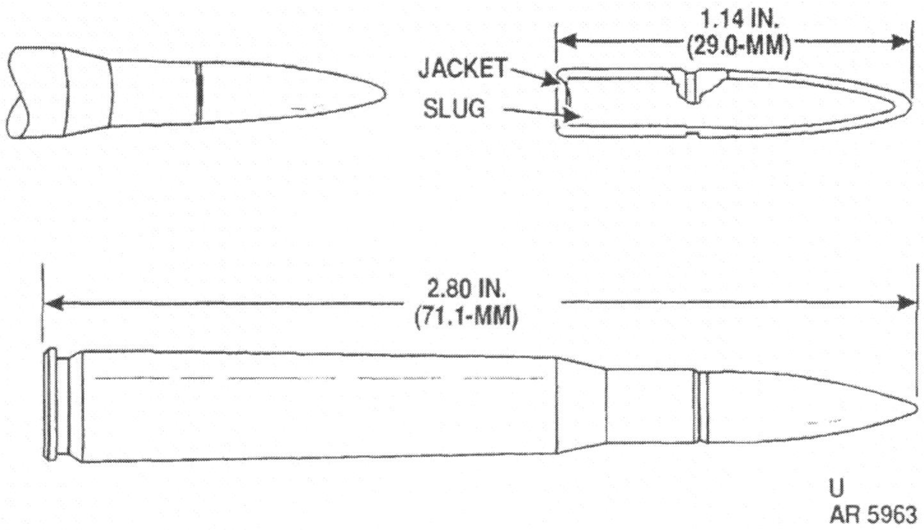


**Type Classification:**

STD - MSR 07798001.

**Use:**

Machine Guns, 7.62mm, M60, M219 and M240; and Rifle, 7.62mm, M14. The cartridge is intended for use against personnel and unarmored targets.

**Description:**

BALL Cartridge. This cartridge is identified by a plain bullet tip.

**Tabulated Data:**

DODAC ........................ 1305-A122
UNO serial number ..... 0012
UNO proper shipping
name ........................ Cartridges for weapons, inert projectile
Weight ........................ 392 gr
Length.......................... 2.8 in. (71.1 mm)
Tracer .......................... NA
Primer ......................... Percussion
Fuze .............. ............. NA

Explosive:
Type ........................ NA
Weight ..................... NA
Incendiary:
Type ........................ NA
Weight ..................... NA
Propellant:
Type ........................ WC 846
Weight ..................... 46 gr
Projectile:
Weight ..................... 146 gr

**Performance:**

Chamber pressure ....... 50,000)() psi
Velocity ....................... 27501) fps, 78 ft from muzzle

**Shipping and Storage Data:**

Quantity-distance class/
SCG ......................... 1.4S
Storage code ............... Class V
DOT shipping class ..... C
DOT designation ......... SMALL ARMS AMMUNITION
Drawing number .......... 10521998

**References:**

SB 700-20
TM 9-1005-223-10
TM 9-1005-223-12P
TM 9-1005-224-10
TM 9-1005-233-10
TM 9-1005-298-12
TM 9-1005-313-10
TM 9-1300-206

## CARTRIDGE, 7.62MM, BALL, M80 (OVERHEAD FIRE APPLICATION)

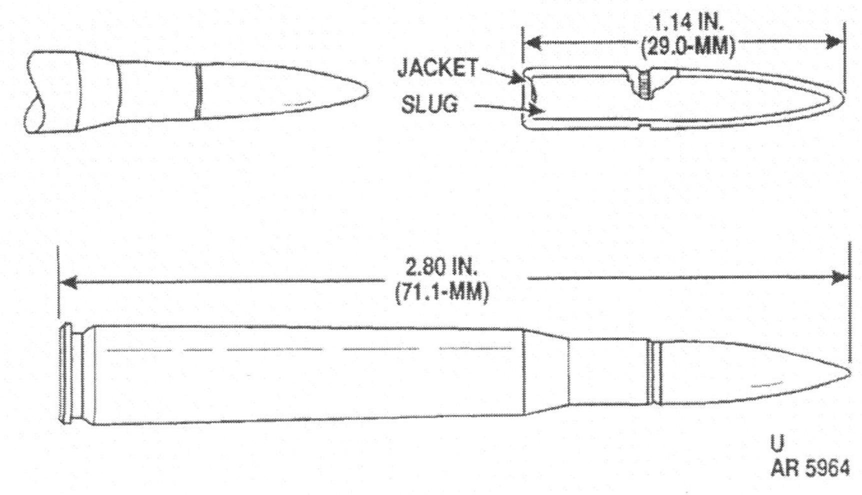


**Type Classification:**

STD - MSR 07798001.

**Use:**

Machine Guns, 7.62mm, M60, M219 and M240. The cartridge is used in machine guns for firing over the heads of troops being trained in field exercises. Stringent production control and screening of ammunition lots ensure the safety of personnel operating immediately below the trajectory of the fired bullets.

**Description:**

BALL Cartridge. The cartridge is identified by a plain bullet tip.

**Tabulated Data:**

DODAC........................ 1305-A166
UNO serial number ..... 0012
UNO proper shipping
name ...................... Cartridges for weapons, inert projectile
Weight ........................ 392 gr
Length ............ ............ 2.8 in. (71.1 mm)
Tracer .......................... NA
Primer ......................... Percussion
Fuze ............................ NA
Explosive:
Type ........................ NA
Weight ...................... NA
Incendiary:
Type ........................ NA
Weight ..................... NA
Propellant:
Type ........................ WC 846
Weight ..................... 46 gr
Projectile:
Weight ..................... 149 gr

**Performance:**

Chamber pressure ....... 50,000 psi
Velocity ....................... 2750 fps, 78 ft from muzzle

**Shipping and Storage Data:**

Quantity-distance class/
SCG ......................... 1.4S
Storage code ............... Class V
DOT shipping class...... C
DOT designation ......... SMALL ARMS AMMUNITION
Drawing number .......... 10523088

**References:**

SB 700-20
TM 9-1005-223-10
TM 9-1005-223-12P
TM 9-1005-224-10
TM 9-1005-233-10
TM 9-1005-298-12
TM 9-1005-313-10
TM 9-1300-206

## CARTRIDGE, 7.62MM, BLANK, M82

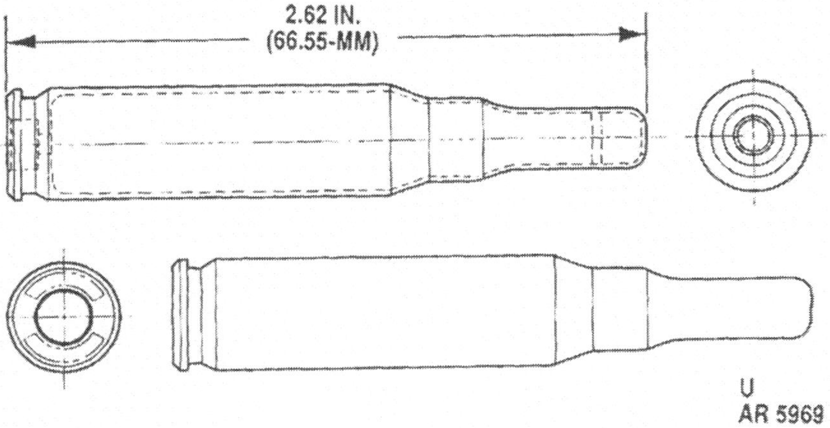


**Type Classification:**

STD - OTCM 36841.

**Use:**

Machine Guns, 7.62mm, M60, M219 and M240; and Rifle, 7.62mm, M14. The cartridge is used in rifles and machine guns equipped with blank firing attachments to simulate firing in training exercises and for saluting purposes.

**Description:**

BLANK Cartridge. The cartridge is identified by its double tapered neck and absence of a bullet.

**Tabulated Data:**

DODAC ........ .............. 1305-A112
UNO serial number ..... 0014
UNO proper shipping
name ....................... Cartridges for weapons, blank
Weight .......... ............. 235 gr
Length ......................... 2.62 in. (66.55 mm)
Tracer .......................... NA
Primer ......................... Percussion
Fuze ............................ NA
Explosive:
Type ........................ NA
Weight ...................... NA
Incendiary:
Type ........................ NA
Weight ...................... NA
Propellant:
Type ........................ SR 8231
Weight ..................... 15 gr

**Performance:**

Chamber pressure ....... NA
Velocity ....................... NA

**Shipping and Storage Data:**

Quantity-distance class/
SCG ............................ 1.4S
Storage code ............... Class V
DOT shipping class ..... C
DOT designation ......... SMALL ARMS AMMUNITION
Drawing number .......... 8597283

**References:**

SB 700-20
TM 9-1005-223-10
TM 9-1005-223-12P
TM 9-1005-224-10
TM 9-1005-233-10
TM 9-1005-298-12
TM 9-1005-313-10
TM 9-1300-206

THIS PAGE INTENTIONALLY LEFT BLANK

## CARTRIDGE, 7.62MM, BALL, SPECIAL, M118

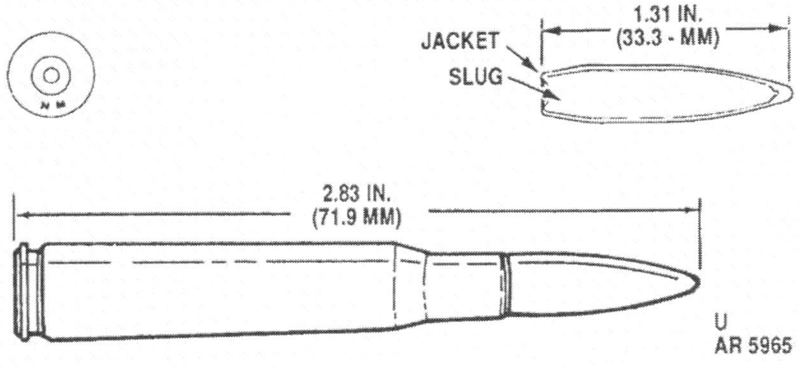


**Type Classification:**

STD - MSR 07798001.

**Use:**

Rifles, 7.62mm, M14, M21, M24, and M40A1. The cartridge is intended and specifically prepared for use in high accuracy weapons.

**Description:**

BALL Cartridge. The cartridge is identified by cartridge case head stampings with NATO design mark, manufacturer and year.

**Tabulated Data:**

DODAC ........................ 1305-A136
UNO serial number ..... 0012
UNO proper shipping
name ....................... Cartridges for weapons, inert projectile
Weight ........................ 390 gr
Length .......................... 2.83 in. (71.9 mm)
Tracer .......................... NA
Primer .......... .............. Percussion
Fuze ............................ NA
Explosive:
Type ........................ NA
Weight ...................... NA
Incendiary:
Type ........................ NA
Weight ..................... NA
Propellant:
Type ........................ WC 846 and IMR 4895
Weight ..................... 44 gr
Projectile:
Weight ...................... 172 gr

**Performance:**

Chamber pressure ....... 50,000 psi
Velocity ....................... 2640 fps

**Shipping and Storage Data:**

Quantity-distance class/
SCG .......................... 1.4S
Storage code ............... Class V
DOT shipping class ..... C
DOT designation ......... SMALL ARMS AMMUNITION
Drawing number .......... 8597555

**References:**

SB 700-20
TM 9-1005-223-12P
TM 9-1300-206

THIS PAGE INTENTIONALLY LEFT BLANK

## CARTRIDGE, 7.62MM, FRANGIBLE, M160

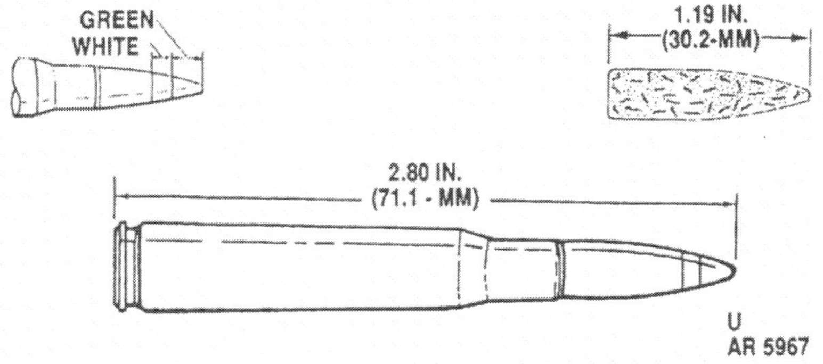


**Type Classification:**

STD - AMCTC 2186.

**Use:**

Machine Guns, 7.62mm, M219 and M240. The cartridge is designed for firing single shots in the machine gun for gunnery practice.

**Description:**

FRANGIBLE Cartridge. The cartridge is identified by a green bullet tip with a white ring to the rear of the green tip.

**Function:**

The frangible bullet, upon striking a target, disintegrates, leaving a mark at the point of impact. It is not to be fired on indoor ranges unless they are well ventilated to prevent the buildup of bullet "dust," which is toxic.

**Tabulated Data:**

DODAC ........................ 1305-A147
UNO serial number ..... 0012
UNO proper shipping
name ....................... Cartridges for weapons, inert projectile
Weight ........................ 315 gr
Length ......................... 2.8 in. (71.1 mm)
Tracer .......................... NA
Primer .......................... Percussion
Fuze ............................ NA
Explosive:
Type .......................... NA
Weight ...................... NA
Incendiary:
Type ........................ NA
Weight ...................... NA
Propellant:
Type ........................ SR 8074
Weight ..................... 10.5 gr
Projectile:
Weight ..................... 108.5 gr

**Performance:**

Chamber pressure ....... NA
Velocity ....................... 1320 fps, 100 yd from muzzle

**Shipping and Storage Data:**

Quantity-distance class/
SCG ......................... 1.4S
Storage code ............... Class V
DOT shipping class ...... C
DOT designation ......... SMALL ARMS AMMUNITION
Drawing number .......... 10522476

**References:**

SB 700-20
TM 9-1005-224-10
TM 9-1005-233-10
TM 9-1005-298-12
TM 9-1005-313-10
TM 9-1300-206

THIS PAGE INTENTIONALLY LEFT BLANK

## CARTRIDGE, DUMMY, 7.62MM, M172

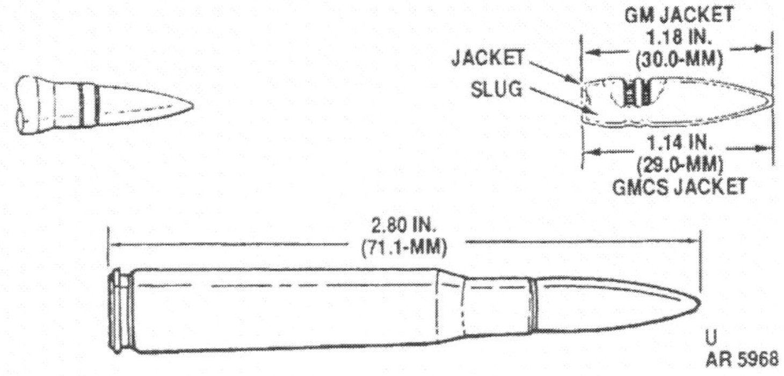


### Type Classification:

STD - AMCTC 4635.

### Use:

Test of 7.62mm weapons. The cartridge is inert and is used to test the mechanism and metallic link belts of 7.62mm weapons.

### Description:

DUMMY Cartridge. The cartridge is identified by a black oxide finish over the entire round and has no primer. There is no vent hole in the primer pocket.

### Tabulated Data:

DODAC........................ 1305-A162
Weight ........................ 385 gr
Length ......................... 2.8 in. (71.1 mm)
Tracer .......... ............... NA
Primer ......................... NA
Fuze ............................ NA
Explosive:
  Type ........................ NA
  Weight ..................... NA
Incendiary:
  Type ......... .............. NA
  Weight ...................... NA
Propellant:
  Type ........................ NA
  Weight ...................... NA
Projectile:
  Weight ...................... NA

### Performance:

Chamber pressure ....... NA
Velocity ...................... NA

### Shipping and Storage Data:

Quantity-distance class/
  SCG ......................... NA
Storage code .............. NA
DOT shipping class ..... NA
DOT designation ......... NONEXPLOSIVE AMMUNITION
Drawing number .......... 10522688

### References:

SB 700-20
TM 9-1005-223-10
TM 9-1005-223-12P
TM 9-1005-224-10
TM 9-1005-233-10
TM 9-1005-298-12
TM 9-1005-313-10
TM 9-1300-206

THIS PAGE INTENTIONALLY LEFT BLANK

## CARTRIDGE, 7.62MM, DIM TRACER, M276

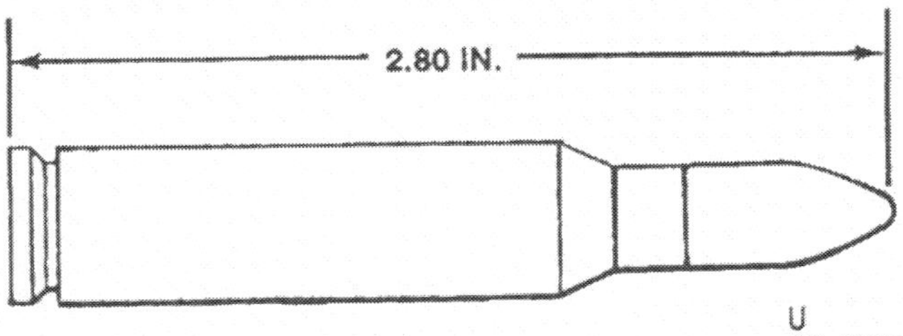


**Type Classification:**

STD - Being performed.

**Use:**

Rifle, 7.62mm, M14; Machine Guns, 7.6mm, M60, M240, M219 and M134. The combat cartridge is to be used by soldiers equipped with night vision equipment.

**Description:**

COMBAT Cartridge. The cartridge is identified by a pink ring behind a green tip.

**Tabulated Data:**

DODAC........................ 1305-A163
UNO serial number ..... 0012
UNO proper shipping
name ........................ Cartridges for weapons, inert projectile
Weight ........................ 381 gr
Length ........................ 2.8 in.
Tracer ........................ R440 Trace Mix
Primer ........................ Percussion
Fuze ........................ NA
Explosive:
Type ........................ NA
Weight ..................... NA
Incendiary:
Type ........................ NA
Weight ...................... NA
Propellant:
Type ........................ WC-846
Weight ..................... 46 gr
Projectile weight .......... 140-150 gr

**Performance:**

Chamber pressure ....... 50,000 psi
Velocity ........................ 2750 fps (GMCS) and 2680 fps (GM), 78 ft from muzzle

**Shipping and Storage Data:**

Quantity-distance class/
SCG ........................ 1.4S
Storage code ............... C
DOT shipping class...... C
DOT designation ......... CARTRIDGES SMALL ARMS

**References:**

DMWR 9-1300-0017-D1
TM 9-1305-201-20&P
TM 9-1305-201-34&P

THIS PAGE INTENTIONALLY LEFT BLANK

## CARTRIDGE, 7.62 MM, MATCH, M852

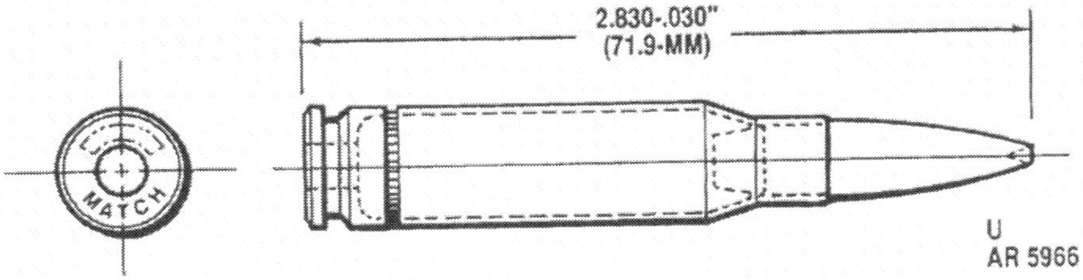


**Type Classification:**
STD - MSR 09816009.

**Use:**

Rifle, 7.62 mm, M14, NATIONAL MATCH. The cartridge is intended and specifically prepared for use in those weapons designated as competitive rifles and also for marksmanship training. The cartridge is not for combat use.

**Description:**

MATCH Cartridge. The cartridge is identified by the cartridge case head stamping of MATCH. It also has a knurl at the base of the cartridge case and a hollow point bullet.

**Tabulated Data:**

DODAC ........................................... 1305-A171
UNO serial number ......................... 0012
UNO proper shipping
  name .......................................... Cartridges for weapons, inert projectile
Weight .......................................... 385 gr
Length ............................................ 2.83 in. (71.9 mm)
Tracer- .............................................. NA
Primer .............................................. Percussion
Fuze ................................................. NA
Explosive:
  Type ............................................. NA
  Weight .......................................... NA
Incendiary:
  Type ............................................. NA
  Weight .......................................... NA
Propellant:
  Type ............................................. IMR 4895
  Weight .......................................... 42 gr
Bullet:
  Type ............................................. Hollow point
  Weight .......................................... 168 gr

**Performance:**

Chamber pressure ............................ 50,000 psi
Velocity ............................................ 2550 fps, 772 nips

**Shipping and Storage Data:**

Quantity-distance class/
SCG .................................................. 1.4S
Storage code .................................... Class V
DOT shipping class ........................... C
DOT designation .............................. SMALL ARMS AMMUNITION
Drawing number ................................ 9341551

THIS PAGE INTENTIONALIY LEFT BLANK

## CARTRIDGE, 7.62 MM, ARMOR PIERCING (AP), M993

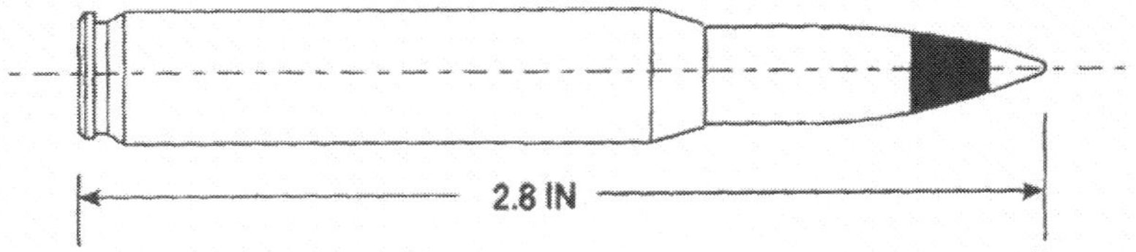


**Type Classification:**

Std - 16 Feb 96.

**Use:**

Rifle, 7.62 MM, M24 Sniper Weapon System Machine Guns, 7.62 mm: M60 and M240.

**Description:**

7.62 mm Armor Piercing Cartridge. The cartridge is identified by a black stripe circling the tip of the projectile. This item will be linked with M62 Tracer.

**Purpose:**

The 7.62 mm cartridge provides an enhanced armor penetrating capability for use in standard 7.62 mm weapons.

**Tabulated Data:**

| | |
|---|---|
| DODAC | 1305-01-398-1450, AA03; 1305-01-398-1449, AA04 |
| Weight | 362.6 gr |
| Length | 2.80 in. |
| Tracer | N/A |
| Primer | Percussion |
| Fuze | N/A |
| Explosive: | |
| Type | N/A |
| Weight | N/A |
| Incendiary: | |
| Type | N/A |
| Weight | N/A |
| Propellant: | |
| Type | Bofors NC1290 |
| Weight | 45 gr |
| Projectile: | |
| Weight | 126.6 gr |

**Performance:**

| | |
|---|---|
| Average chamber pressure at 70°F | 55,115 psi |
| Average velocity at 70°F | 2985 fps at 78 ft from muzzle |

**Shipping and Storage Data:**

| | |
|---|---|
| M993 AP, carton packed | 1305-01-398-1450, DODAC AA03 |
| M993 AP and M62 tracer, linked in 24:1 ratio | 1305-01-398-1449, DODAC AA04 |
| Quantity-distance class/SCG | 1.4C |
| Storage code | Class C |
| DOT shipping class | C |
| DOT designation | SMALL ARMS AMMUNITION |
| Drawing number | 12956134 |

**References:**

TM 9-1005-224-10
TM 9-1005-306-10
TM 9-1005-313-10
TM 9-1305-201-20&P
TM 9-1305-201-34&P

THIS PAGE INTENTIONALLY LEFT BLANK



☆U.S. GOVERNMENT PRINTING OFFICE: 1996-545-437/40708

# CHAPTER 12

# 9 MM CARTRIDGES

THIS PAGE INTENTIONALLY LEFT BLANK

## CARTRIDGE, 9 MM, BALL

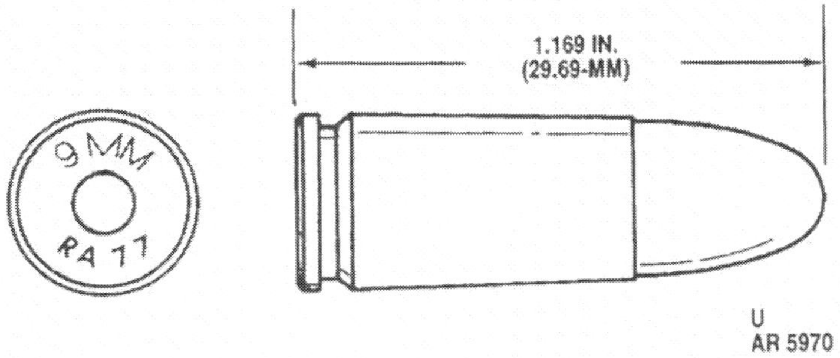


**Type Classification:**

Commercial.

**Use:**

Modified M3 Submachine Gun or commercial weapons. (The use in M9 Pistol is not authorized.) The cartridge is intended for use against personnel.

**Description:**

BALL Cartridge.

**Tabulated Data:**

| | |
|---|---|
| DODAC | 1305-A360 |
| UNO serial number | 0012 |
| UNO proper shipping name | Cartridges for weapons, inert projectile |
| Weight | 182 gr |
| Length | 1.169 in. (29.69 mm) |
| Tracer | NA |
| Primer | Percussion |
| Fuze | NA |
| Explosive: | |
| Type | NA |
| Weight | NA |
| Incendiary: | |
| Type | NA |
| Weight | NA |
| Propellant: | |
| Type | Commercial |
| Weight | MBR |
| Projectile weight | 115 gr |

**Performance:**

**NOTE**
**Mid case pressures normally run 8000 to 10,000 psi higher than case mouth pressure.**

| | |
|---|---|
| Mid case pressure | 38,500 psi (avg), 43,000 psi (max) |
| Velocity | 1125 + 90 fps, 15 ft from muzzle |

**Shipping and Storage Data:**

| | |
|---|---|
| Quantity-distance class/ SCG | 1.4S |
| Storage code | Class V |
| DOT shipping class | C |
| DOT designation | SMALL ARMS AMMUNITION |
| Drawing number | 10542448 |

**References:**

SB 700-20
TM 9-1300-206

THIS PAGE INTENTIONALLY LEFT BLANK

**CARTRIDGE, 9 MM, BALL, NATO, M882**

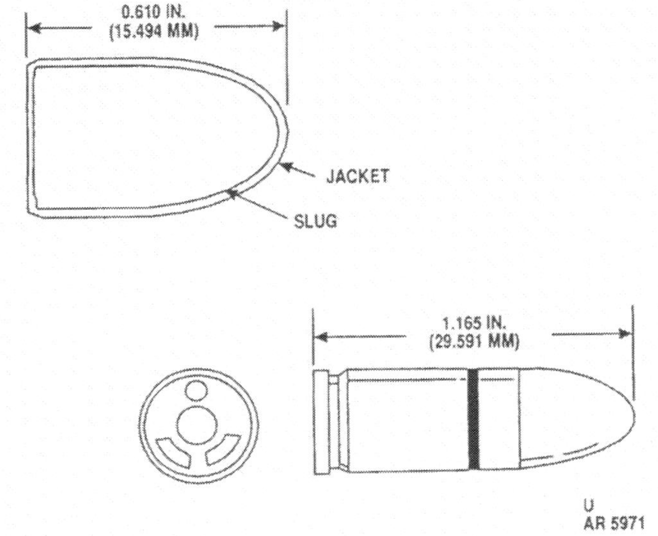


**Type Classification:**

HQDA (DAMA-CSM), dated 15 April 85.

**Use:**

Pistol, 9 mm, M9. The cartridge is intended for use against personnel.

**NOTE**

**M882 ammunition can be used in the following nonstandard weapons:**

| Pistol | Submachine Guns |
|---|---|
| HK P7 series (GE) | HK MP5 (GE) |
| Walther P38 (GE) | HK MP5SD (GE) |
| FN P35 (BE) | IMI Uzi (IS) |
| | Baretta M12 (IT) |
| | FFV M45/45B (SW) |
| | Madsen M50 (DE) |
| | CZ M23 (CZ) |
| | Sten MK II (UK) |
| | Sterling 12A3 (UK) |

**Description:**

BALL Cartridge. The cartridge is identified by the plain bullet tip.

**Tabulated Data**

| | |
|---|---|
| DODAC | 1305-A3(3 |
| UNO serial number | 0012 |
| UNO proper shipping name | Cartridges for weapons, inert projectile |
| Weight | 179 gr |
| Length | 1.165 in. (29.591 mm) |
| Tracer | NA |
| Primer | Percussion |
| Fuze | NA |
| Explosive: | |
| Type | NA |
| Weight | NA |
| Incendiary: | |
| Type | NA |
| Weight | NA |
| Propellant: | |
| Type | HPC 26 |
| Weight | 6 gr |
| Projectile weight | 112 gr |

**Performance:**

Case mouth pressure ........................31,175 psi (avg), 36,250 psi (max)
Velocity ............................................1263 ± 5 fps, 15 ft from muzzle

**Shipping and Storage Data:**

Quantity-distance class/ SCG .............................................1.4S
Storage code .................................... Class V
DOT shipping class ..........................C
DOT designation ..............................SMALL ARMS AMMUNITION
Drawing number................................9345211

**References:**

SB 700-20
TM 9-1300-206

## CARTRIDGE, 9MM, HIGH PRESSURE TEST, M905

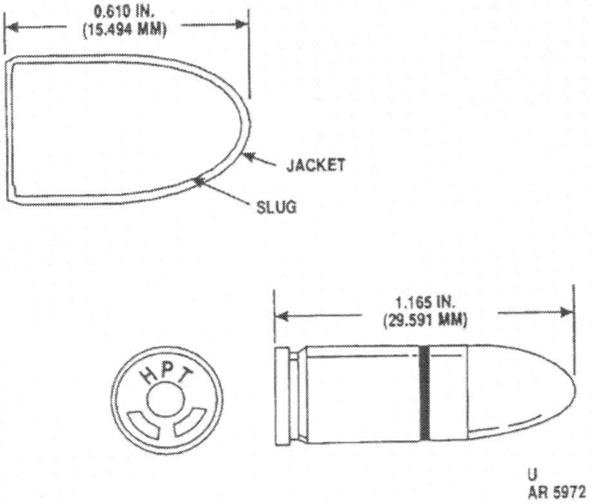


**Type Classification:**

HQDA (DAMA-CSM), dated 15 April 85.

**Use:**

Proof testing of M9 pistols. The cartridge is intended for use in proofing new pistols and test barrels to demonstrate safety prior to releasing weapons to the field.

**Description:**

HIGH PRESSURE TEST Cartridge. The cartridge is identified by a tinned cartridge case and high pressure test head stamp.

**Tabulated Data:**

DODAC ............................................1305-A364
UNO serial number ........................... 0012
UNO proper shipping
name................................................ Cartridges for weapons, inert projectile
Weight .............................................. 179 gr
Length.............................................. 1.165 in. (29.591 mm)
Tracer ............................................... NA
Primer............................................... Percussion
Fuze.................................................. NA
Explosive:
Type ............................................ NA
Weight..........................................NA
Incendiary:
Type ............................................NA
Weight .........................................NA
Propellant:
Type ............................................. WC 370
Weight..........................................7.5 gr

**Performance:**

Chamber pressure ............................50,000 psi
Velocity ............................................ NA

**Shipping and Storage Data:**

Quantity-distance class/
SCG ............................................. 1.4C
Storage code.................................... Class V
DOT shipping class ........................... C
DOT designation ............................... SMALL ARMS AMMUNITION
Drawing number ..............................9378265

**References:**

SB 700-20
TM 9-1300-206

THIS PAGE INTENTIONALLY LEFT BLANK

# CHAPTER 13

## 14.5 MM CARTRIDGES

THIS PAGE INTENTIONALLY LEFT BLANK

## CARTRIDGE, 14.5 MM, TRAINER-SPOTTER, M181A1

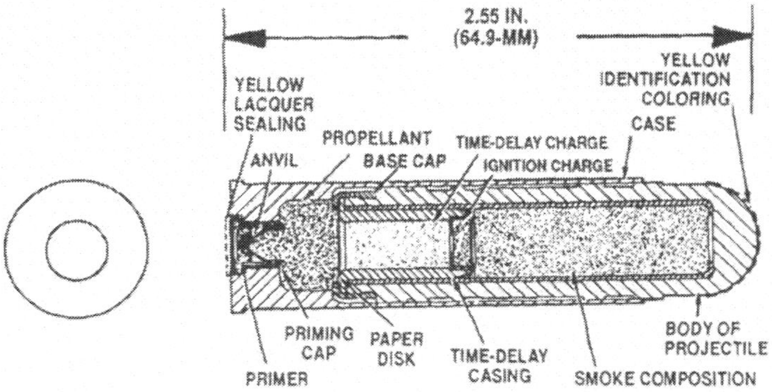


**Type Classification:**

STD - AMCTC 138.

**Use:**

Artillery Trainer, M31.

**Description:**

TRAINER-SPOTTER Cartridge. The cartridge has a chemical-filled (smoke composition) projectile. It is identified by the markings on the cartridge case and a yellow projectile tip. The cartridge has no fuze and it uses a time delay element to provide an air burst.

**Function:**

When the cartridge is fired, the heat of the propellant gases ignites the delay element. The delay element burns 3 seconds and, in turn, ignites the smoke composition. The heat and pressure produced by the burning filler rupture the plastic closing disc at the projectile base and a visible puff of smoke is emitted.

**Tabulated Data:**

DODAC ............................................1305-A365
UNO serial number ..........................0012
UNO proper shipping
name ........................................... Cartridges for weapons, inert projectile

Weight ............................................1065 gr
Length ............................................2.55 in. (64.9 mm)
Tracer ............................................NA
Primer ............................................Percussion
Fuze ............................................Powder train delay
Explosive:
Type ............................................NA
Weight ............................................NA
Incendiary:
Type ............................................NA
Weight ............................................NA
Propellant:
Type ............................................Hi Skor 700X
Weight ............................................3.5 gr

**Performance:**

Chamber pressure............................. NA
Velocity ............................................100 mps, 4 m from muzzle; 328 fps, 15 ft from muzzle

**Shipping and Storage Data:**

Quantity-distance class/
SCG ............................................1.4S
Storage code.................................... Class V
DOT shipping class ...........................C
DOT designation ............................... SMALL ARMS AMMUNITION
Drawing number................................ 9098295

**References:**

SB 700-20
TM 9-1300-206

THIS PAGE INTENTIONALLY LEFT BLANK

## CARTRIDGE, 14.5 MM, TRAINER-SPOTTER, M182A1

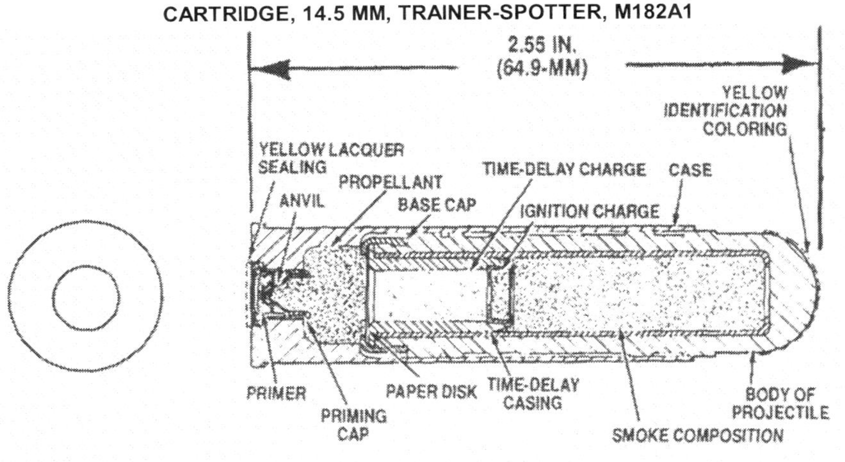


**Type Classification:**

STD - AMCTC 138.

**Use:**

Artillery Trainer, M31.

**Description:**

TRAINER-SPOTTER Cartridge. The cartridge has a chemical-filled (smoke composition) projectile. It is identified by the markings on the cartridge case and a yellow projectile tip. The cartridge has no fuze and it uses a time delay element to provide an air burst.

**Function:**

When the cartridge is fired, the heat of the propellant gases ignites the delay element. The delay element burns 6 seconds and, in turn, ignites the smoke composition. The heat and pressure produced by the burning filler rupture the plastic closing disc at the projectile base and a visible puff of smoke is emitted.

**Tabulated Data:**

DODAC ............................................1305-A366
UNO serial number ..........................0012
UNO proper shipping
name ...............................................Cartridges for weapons, inert projectile

Weight ...................................... 1065 gr
Length...................................... 2.55 in. (64.9 mm)
Tracer- ...................................... NA
Primer...................................... Percussion
Fuze ...................................... Powder train delay
Explosive:
Type - ...................................... NA
Weight ...................................... NA
Incendiary:
Type ...................................... NA
Weight...................................... NA
Propellant:
Type ...................................... Hi Skor 700X
Weight-...................................... 3.5 gr
Performance:

Chamber pressure...................... NA
Velocity ...................................... 100 ips, 4 m from muzzle; 328 fps, 15 ft from muzzle

**Shipping and Storage Data:**

Quantity-distance class/
SCG ...................................... 1.4S
Storage code ............................. Class V
DOT shipping class .................... C
DOT designation ........................ SMALL ARMS AMMUNITION
Drawing number ........................ 9098297

**References:**

SB 700-20
TM 9-1300-206

THIS PAGE INTENTIONALLY LEFT BLANK

## CARTRIDGE, 14.5 MM, TRAINER-SPOTTER, M183A1

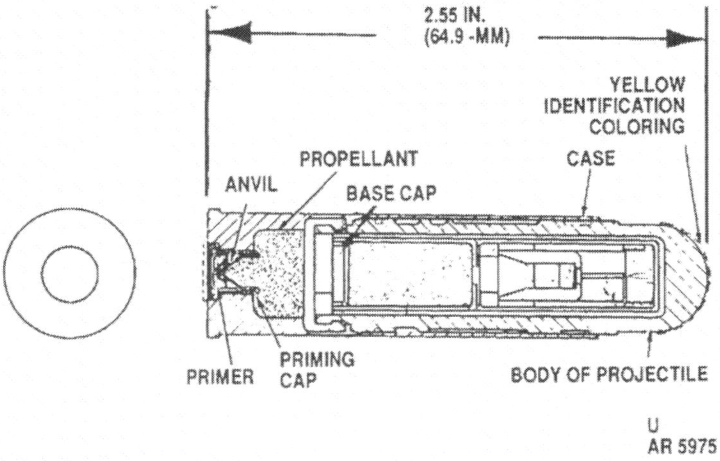


### Type Classification:

STD - AMCTC 138.

### Use:

Artillery Trainer, M31.

### Description:

TRAINER-SPOTTER Cartridge. The cartridge consists of a cartridge case, a chemical filled (smoke composition) projectile, and a fuze with a small explosive charge to provide a visible and audible verification at point of impact. The cartridge is identified by the markings on the cartridge case and a yellow projectile tip.

### Function:

The projectile impacts with the target. The point detonating fuze functions and an explosion charge within the fuze ignites the smoke composition and ejects it out the base of the projectile.

### Tabulated Data:

| | |
|---|---|
| DODAC | 1305-A367 |
| UNO serial number | 0012 |
| UNO proper shipping name | Cartridges for weapons, inert projectile |
| Weight | 1065 gr |
| Length | 2.55 in. (64.9 mm) |
| Tracer | NA |
| Primer | Percussion |
| Fuze | PD |
| Explosive: | |
| Type | NA |
| Weight | NA |
| Incendiary: | |
| Type | NA |
| Weight | NA |
| Propellant: | |
| Type | Hi Skor 700X |
| Weight | 3.5 gr |
| Performance: | |
| Chamber pressure | NA |
| Velocity | 100 nips, 4 m from muzzle; 328 fps, 15 ft from muzzle |

### Shipping and Storage Data:

| | |
|---|---|
| Quantity-distance class/ SCG | 1.4S |
| Storage code | Class V |
| DOT shipping class | C |
| DOT designation | SMALL ARMS AMMUNITION |
| Drawing number | 9098293 |

### References:

SB 700-20
TM 9-1300-206

THIS PAGE INTENTIONALLY LEFT BLANK

# CHAPTER 14

# 20MM CARTRIDGES

THIS PAGE INTENTIONALLY LEFT BLANK

## CARTRIDGE, DUMMY, 20MM, M51A2

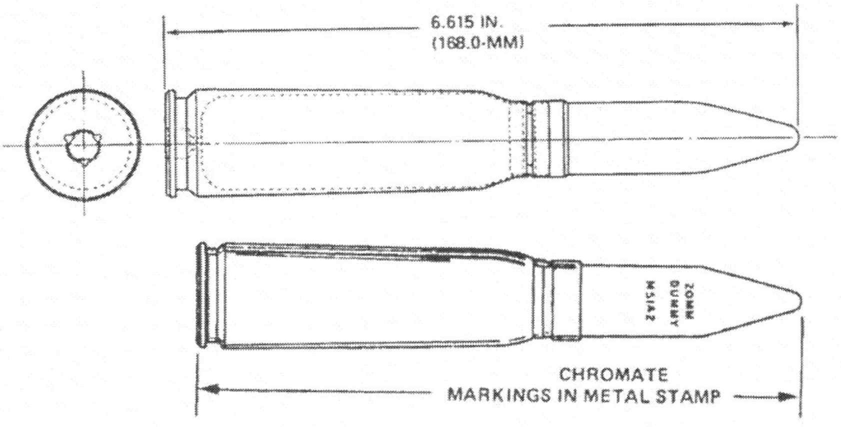


**Type Classification:**

STD - AMCTC 6900.

**Use:**

Guns. 20mm, M39, M61, M139, M168, M195 and M197. The cartridge is a completely inert assembly used for filling the conveyor system during shipment or storage and for nonfiring system checkout.

**Description:**

DUMMY Cartridge.

**Tabulated Data:**

DODAC........................................1305-A924
Weight ........................................ 3850 gr
Length..........................................6.615 in. (168 mm)
Tracer .......................................... NA
Primer ..........................................NA
Fuze.............................................NA
Explosive:
Type ............................................NA
Weight .........................................NA

Incendiary:
Type ............................................NA
Weight .........................................NA
Propellant:
Type ............................................NA
Weight..........................................NA

**Performance:**

Chamber pressure ............................NA
Velocity ............................................NA

**Shipping and Storage Data:**

Quantity-distance class/
SCG .............................................NA
Storage code ....................................NA
DOT shipping class ..........................NA
DOT designation .............................. NONEXPLOSIVE AMMUNITION
Drawing number ...............................11075758

**References:**

SB 700-20
TM 9-1005-286-10
TM 9-1300-206

THIS PAGE INTENTIONALLY LEFT BLANK

## CARTRIDGE, 20 MM, ARMOR PIERCING INCENDIARY-TRACER, M52R1

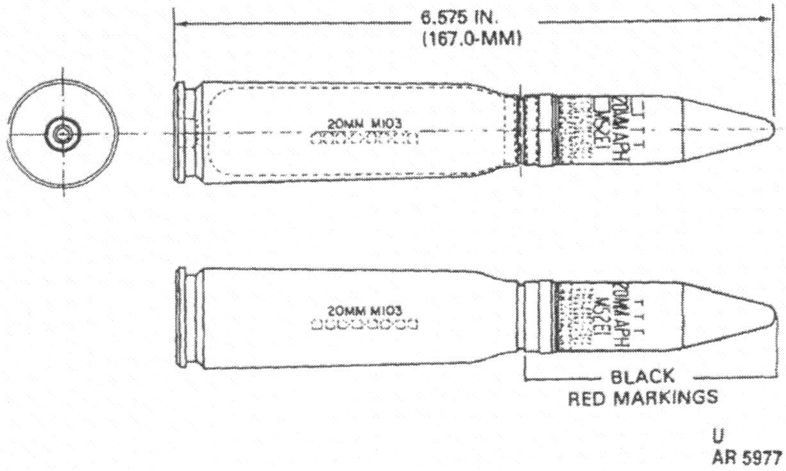


**Type Classification:**

STD - OI'CM 37404.

**Use:**

Guns, 20mm, M39, M61 and M195. The cartridge is used against moderately armed targets and when incendiary e'fect is needed.

**Description:**

ARMCR PIERCING INCENDIARY-TRACER Cartridge. The armor piercing shot is the main part of the projectile. The ballsitic cap (ogive) contains an incendiary composition. The cartridge is identified by a projectile painted black with red markings.

**Function:**

The ncse/ogive of the projectile impacts with the target and ruptures, and the incendiary composition ignites upon exposure to the air.

**Tabulated Data:**

| | |
|---|---|
| DODAC | 1305-A870 |
| Weight | 3900 gr |
| Length | 6.575 in. (167 mm) |
| Tracer- | R284 |
| Primer | Electric, M52A3B1 |
| Fuze | NA |
| Explosive: | |
| Type | NA |
| Weight- | NA |
| Incendiary: | |
| Type | 1280 |
| Weight | 65 gr |
| Propellant: | |
| Type | WC 870 or WC 872 |
| Weight- | MBR (585-805) |

**Performance:**

| | |
|---|---|
| Chamber pressure | 60,500 psi |
| Velocity | 3380 fps, 78 ft from muzzle |

**Shipping and Storage Data:**

| | |
|---|---|
| Quantity-distance class/ SCG | 1.4G |
| Storage code | Class V |
| DOT shipping class | A |
| DOT designation | AMMUNITION FOR SMALL ARMS WITH INCENDIARY PROJECTILES |
| Drawing number | 7259218 |

**References:**

SB 700-20
TM 9-1300-206

THIS PAGE INTENTIONALLY, LEFT BLANK

CARTRIDGE, 20 MM, ARMOR PIERCING INCENDIARY, M53

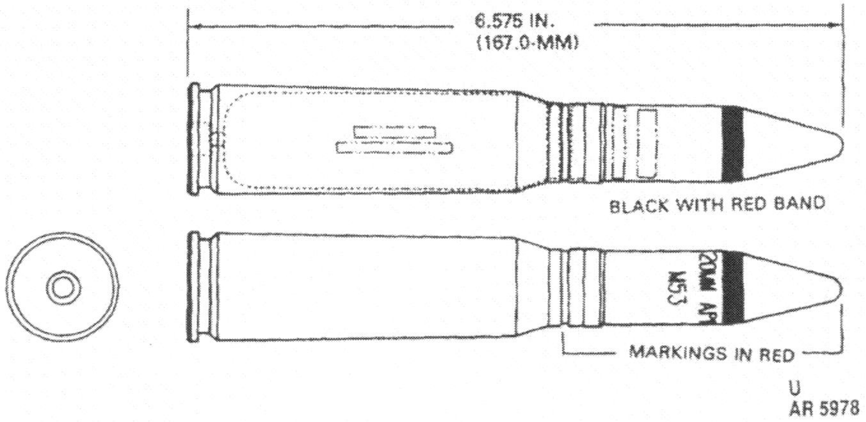


**Type Classification:**

STD - OTCM 36841.

**Use:**

Gulls, 20mm, M39, M61 and M168. The cartridge is used against armored targets, and functions with a combined incendiary and penetrating effect.

**Description:**

ARMOR PIERCING INCENDIARY Cartridge. The body of the projectile is a solid shot made from bar alloy steel. The nose of the projectile is made from aluminum alloy and is charged with three increments of incendiary composition. The projectile is painted black with red band and red markings.

**Tabulated Data:**

DODAC........ ............................1305-A798
Weight ...................................... 3980 gr
Length....................................... 6.575 in. (167 mm)
Tracer .......................................NA
Primer .....................................Electric, M52A3B
Fuze.......................................... NA
Explosive:
Type .......... ........................... NA
Weight.................................. NA
Incendiary:
Type ...................................... No. 23 or No. 136 and No. 68
Weight.................................. 65 gr
Propellant:
Type ...................................... WC 870 or WC 872
Weight .................................. MBR (585-605 gr)

**Performance:**

Chamber pressure ..................... 60,500 psi
Velocity .................................... 3380 fps, 50 ft from muzzle

**Shipping and Storage Data:**

Quantity-distance class/
SCG ..................................... 1.4G
Storage code............................. Class V
DOT shipping class .................... A
DOT designation ....................... AMMUNITION FOR SMALL ARMS WITH INCENDIARY PROJECTILES
Drawing number......................... 7258802

**References:**

SB 700-20
TM 9-1300-206

THIS PAGE INTENTIONALLY LEFT BLANK

## CARTRIDGE, 20 MM, BALI,, HIGH PRESSURE TEST, M54A1

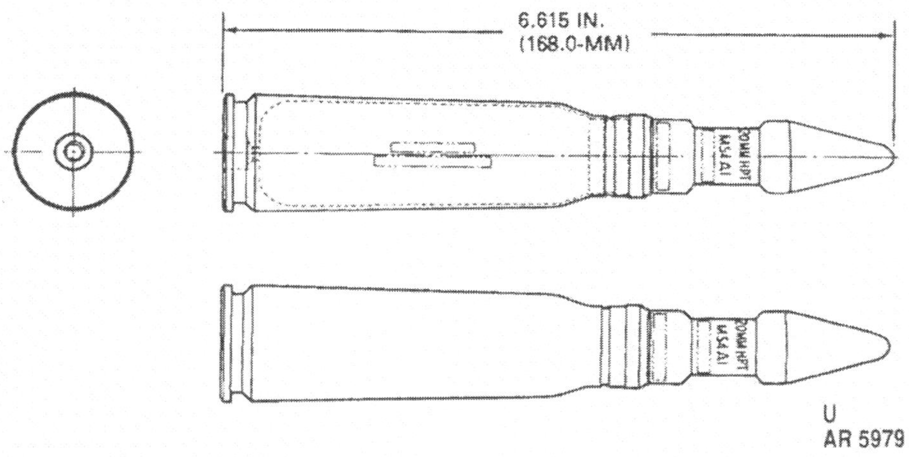


**Type Classification:**

STD - AMCTC 6899.

**Use:**

Guns, Automatic, 20mm, M39 and M61 chambered for the M103 case configuration. The cartridge is used for proof testing of 20mnl weapons and barrels during manufacture and repair.

**Description:**

BALL Cartridge. The cartridge is a solid steel projectile, under cut behind the ogive to control projectile weight, for developing high chamber pressure. The cartridge is identified by a stannic-stained (silvered) cartridge case.

**Tabulated Data:**

DODAC .................................... 1305-A892
UNO serial number .................... 0339
UNO proper shipping
name ........................................ Cartridges for weapons, inert projectiles
Weight .................................... 4392 gr
Length....................................... 6.615 in. (168 mm)
Tracer ....................................... NA
Primer....................................... Electric, M52A3B1
Fuze ........................................... NA
Explosive:
Type ........................................ NA
Weight .................................... NA
Incendiary:
Type ........................................ NA
Weight..................................... NA
Propellant:
Type ......................................... WC 870 or WC 872
Weight .................................... MBR

**Performance:**

Chamber pressure........................ 62,500 to 72,500 psi (copper)
Velocity ...................................... NA

**Shipping and Storage Data:**

Explosive hazard class................. 1
Quantity-distance class/
SCG ....................................... 1.4C
Storage code ..............................Class V
DOT shipping class ......................C
DOT designation .........................SMALL ARMS AMMUNITION
Drawing number .........................7258858

**References:**

SB 700-20
TM 9-1300-206

THIS PAGE INTIENTIONALLY LEFT BLANK

## CARTRIDGE, 20MM, TARGET PRACTICE, M55A2

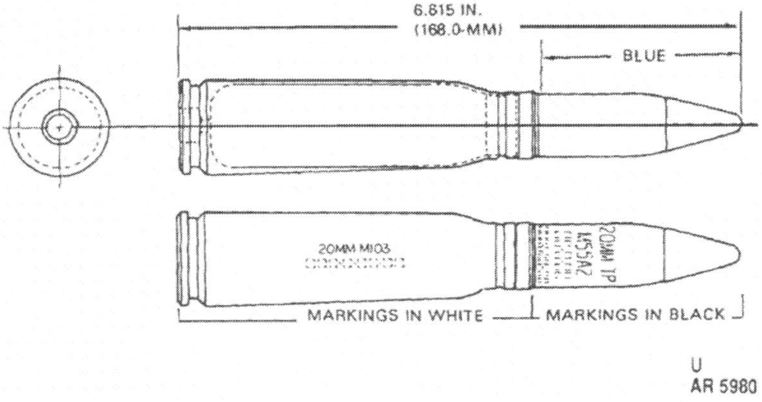


**Type Classification:**

STD - AMCTC 7741.

**Use:**

Guns, 20mm, M39, M61 and M68. The cartridge is a training round intended as a substitute for a service round to be used in target practice.

**Description:**

TARGET PRACTICE Cartridge. The one piece projectile cavity is hollow. The nose (ogive) made of aluminum or steel is swaged to the projectile body. The projectile is painted blue with markings in black. It consists of a solid projectile without a fuze.

**Function:**

The projectile is inert, therefore no specific function takes place upon impact except penetration of the target.

**Tabulated Data:**

| | |
|---|---|
| DODAC | 1305-A897 |
| UNO serial number | 0339 |
| UNO proper shipping name | Cartridges for weapons, inert projectiles |
| Weight | 3935 gr |
| Length | 6.615 in. (168 mm) |
| Tracer | NA |
| Primer | Electric, M52A3B1 |
| Fuze | NA |
| Explosive: | |
| Type | NA |
| Weight | NA |
| Incendiary: | |
| Type | NA |
| Weight | NA |
| Propellant: | |
| Type | WC 870 or WC 872 |
| Weight | MBR (585 or 605 gr) |

**Performance:**

| | |
|---|---|
| Chamber pressure | 60,500 psi |
| Velocity | 3380 fps, 78 ft from muzzle |

**Shipping and Storage Data:**

| | |
|---|---|
| Quantity-distance class/ SCG | 1.4C |
| Storage code | Class V |
| DOT shipping class | C |
| DOT designation | SMALL ARMS AMMUNITION |
| Drawing number | 7258817 |

**References:**

SB 700-20
TM 9-1005-286-10
TM 9-1300-206

THIS PAGE INTENTIONALLY LEFT BLANK

## CARTRIDGE, 20MM, HIGH EXPLOSIVE INCENDIARY, M56A3/M56A4

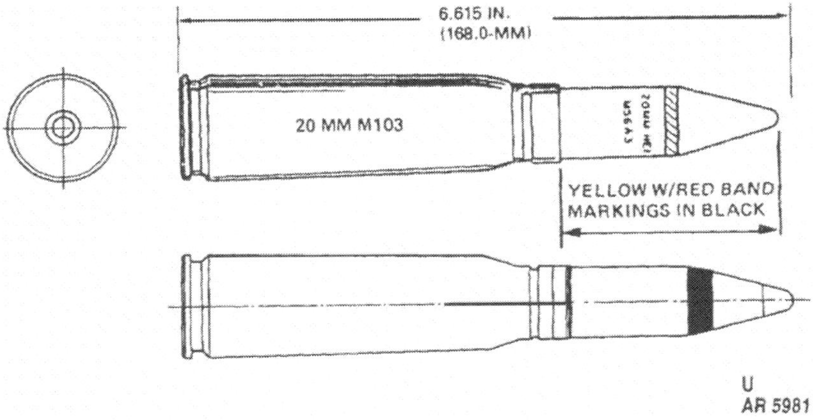


**Type Classification:**

M56A3: STD - LCC-B, MSR 03816015.
M56A4: STD - LCC-A, MSR 03816015.

**Use:**

Guns, 20mm, M39, M61, M168 and M195. The cartridge is for use against ground targets, including lightly armored vehicles, functioning with both explosive and incendiary effect. The projectile consists of a high explosive incendiary (HEI) charge and is assembled with the M505A3 fuze.

**Description:**

HIGH EXPLOSIVE INCENDIARY Cartridge. The projectile is thin-walled steel. A base plate is attached to the projectile to prevent ignition of the incendiary mixture by the propellant gases. A point-detonating (PD) fuze is screwed into the nose of the projectile.

**Function:**

The M505A3 fuze is a point-detonating, single-action fuze intended to function upon impact with the target.

**Difference Between Models:**

The models differ in the method of loading the projectiles. The M56A3 has the HE mix and the incendiary mix combined in one pellet. The M56A4 has the incendiary pellet inserted into the projectile, then the HE mix pellet is added. This improves fire start capability.

**Tabulated Data:**

| | |
|---|---|
| DODAC | 1305-A890 |
| UNO serial number | 0321 |
| UNO proper shipping name | Cartridges for weapons, with bursting charge |
| Weight | 3965 gr |
| Length | 6.615 in. (168 mm) |
| Tracer | NA |
| Primer | Electric, M52A3B1 |
| Fuze | PD, M505A3 |
| Explosive: | |
| Type | H761 |
| Weight | 165 gr |
| Incendiary: | |
| Type | I136 |
| Weight | 20 gr |
| Propellant: | |
| Type | WC 870 or WC 872 |
| Weight | MBR (585 or 605 gr) |

**Performance:**

Chamber pressure........................ 60,500 psi
Velocity........................................ 3380 fps, 78 ft from muzzle

**Shipping and Storage Data:**

Quantity-distance class/
SCG........................................... 1.2E (04)
Storage code................................ Class V
DOT shipping class...................... A
DOT designation.......................... AMMUNITION FOR SMALL ARMS WITH EXPLOSIVE PROJECTILES
Drawing number........................... 7259023 (M56A3); 11075861 (M56A4)

**References:**

DARCOM 700-3-2
SB 700-20
TM 9-1005-286-10
TM 9-1300-206

## CARTRIDGE, 20MM, ARMOR PIERCING-TRACER, M95

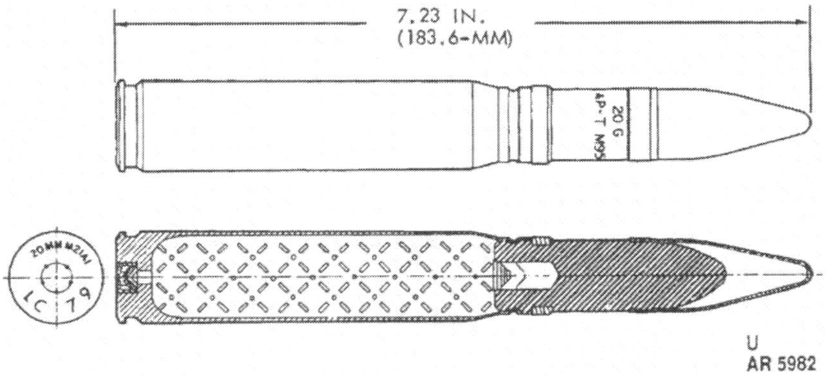


**Type Classification:**

CON - OTCM 36841.

**Use:**

Guns, 20mm, M3 and M24A1.

**Description:**

ARMOR PIERCING-TRACER Cartridge. The projectile is a solid shot type, made from bar or forged steel. A steel windshield is crimped to the forward part of the projectile body. The base of the projectile contains a red tracer composition.

**Function:**

When the primer is initiated, the propellant ignites. Ignition of the propellant ignites the tracer composition in the base of the projectile. The tracer burns the first 2.25 seconds of projectile flight, or an approximate range of 1,400 yards.

**Tabulated Data:**

DODAC........................................ 1305-A848
Weight ........................................ 3990 gr
Length........................................ 7.23 in. (183.6 mm)
Tracer ........................................ R284
Primer........................................ Percussion
Fuze........................................ NA
Explosive:
Type ........................................ NA
Weight ........................................ NA
Incendiary:
Type ........................................ NA
Weight ........................................ NA
Propellant:
Type ........................................ 75-1-177A
Weight ........................................ 470 gr

Performance:

Chamber pressure........................ 51,000 psi
Velocity ........................................ 2680 fps, 78 ft from muzzle

**Shipping and Storage Data:**

Quantity-distance class/
SCG........................................ 1.4C
Storage code................................ Class V
DOT shipping class ...................... C
DOT designation .......................... SMALL ARMS AMMUNITION
Drawing number........................... 75-1-177

**References:**

SB 700-20
TM 9-1300-206

THIS PAGE INTENTIONALLY LEFT BLANK

## CART'RIDGE, 20MM, INCENDIARY, M96

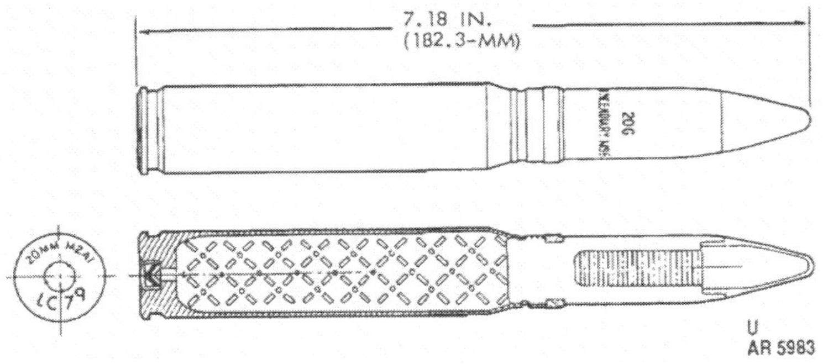


**Type Classification:**

CON - OTCM 36841.

**Use:**

Guns, 20mm, M3 and M24A1.

**Description:**

INCENDIARY Cartridge. The body of the projectile is made from cold dracon steel. The projectile nose and body are filled with incendiary composition.

**Function:**

Impact with the target by the projectile produces compression of the heat and friction sensitive elements in the incendiary mixture causing ignition of the mixture.

**Tabulated Data:**

DODAC ........................................ 1305-A776
UNO serial number ...................... 0009
UNO proper shipping
  name ........................................ Ammunition, incendiary

Weight ........................................ 3850 gr
Length ........................................ 7.18 in. (182.3 mm)
Tracer ........................................ NA
Primer ........................................ Percussion
Fuze ........................................ NA
Explosive:
  Type ........................................ NA
  Weight ........................................ NA
Incendiary:
  Type ........................................ I136
  Weight ........................................ 3.92 oz
Propellant:
  Type ........................................ 75-1-178A
  Weight ........................................ 470 gr

**Performance:**

Chamber pressure ........................ 51,000 psi
Velocity ........................................ 2680 fps, 78 ft from muzzle

**Shipping and Storage Data:**

Quantity-distance class/
  SCG ........................................ 1.2G
Storage code ................................ Class V
DOT shipping class ...................... A

DOT designation .......................... AMMUNITION FOR SMALL ARMS WITH INCENDIARY PROJECITLES
Drawing number........................... 75-1-178

**References:**

SB 700-20
TM 9-1300-206

## CARTRIDGE, 20MM, HIGH EXPLOSIVE INCENDIARY, M97A2

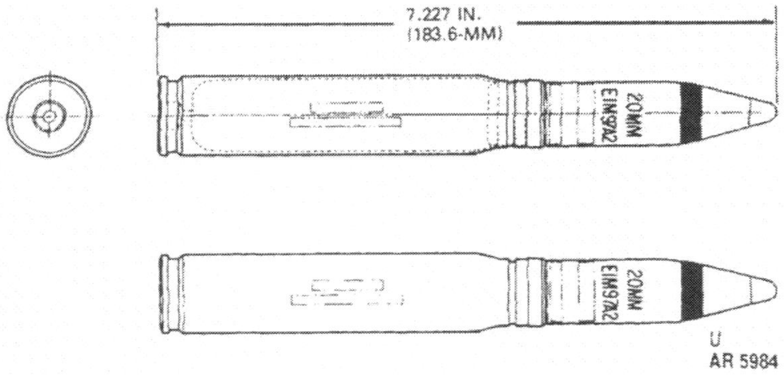


**Type Classification:**

STD - AMCTC 4376.

**Use:**

Gun, 20mm, M24A1.

**Description:**

HIGH EXPLOSIVE INCENDIARY Cartridge. The projectile has a hollow body filled with an incendiary composition and a high explosive (called RDX). The base of the projectile has a base cover and the forward end has a point-detonating (PD) fuze.

**Function:**

When the primer is initiated, the propellant ignites. Ignition of the propellant provides hot gases to propel the projectile from the weapon. Upon impact, the point detonating fuze functions, initiating the high explosive filler and dispersing the incendiary composition.

**Tabulated Data:**

| | |
|---|---|
| DODAC | 1305-A845 |
| Weight | 4000 gr |
| Length | 7.227 in. (183.6 mm) |
| Tracer | NA |
| Primer | Electric, M52A3B1 |
| Fuze | PD, M505A3 |
| Explosive: | |
| Type | H761 |
| Weight | 120 gr |
| Incendiary: | |
| Type | NA |
| Weight | NA |
| Propellant: | |
| Type | WC 875 or IMR 7013 |
| Weight | MBR |

**Performance:**

| | |
|---|---|
| Chamber pressure | 51,000 psi |
| Velocity | 2680 fps, 78 ft from muzzle |

**Shipping and Storage Data:**

| | |
|---|---|
| Quantity-distance class/ SCG | 1.2E |
| Storage code | Class V |
| DOT shipping class | A |
| DOT designation | AMMUNITION FOR SMALL ARMS WITH EXPLOSIVE PROJECTILE |
| Drawing number | 7259063 |

**References:**

SB 700-20
TM 9-1300-206

THIS PAGE INTENTIONALLY LEFT BLANK

## CARTRIDGE, 20MM, TARGET PRACTICE, M99A1

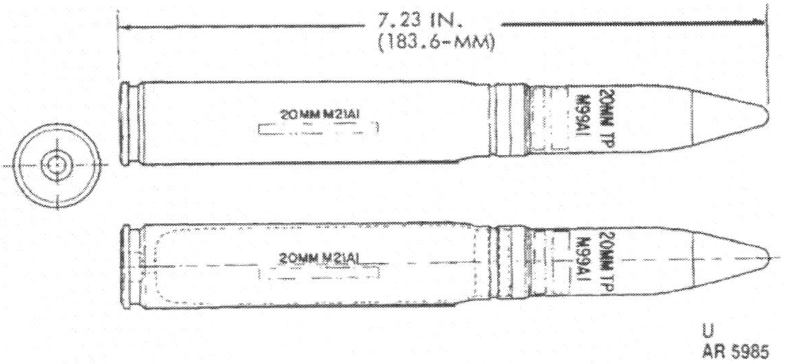


**Type Classification:**

STD - OTCM 37403.

**Use:**

Gun, 20rmm, M24A1. The cartridge is used for training, gun testing, and gull system function tests.

**Description:**

TARGET PRACTICE Cartridge. The cartridge contains a hollow, inert projectile which has a cavity to control the weight. The aluminum nose is swaged to the body.

**Function:**

The projectile is inert, therefore no specific function takes place except target penetration.

**Tabulated Data:**

DODAC........................................ 1305-A847
Weight........................................ 4000 gr
Length........................................ 7.23 in. (183.6 mm)
Trace........................................ NA
Primer........................................ Electric, M52A3B1
Fuze........................................ NA

Explosive:
Type........................................ NA
Weight........................................ NA
Incendiary:
Type........................................ NA
Weight........................................ NA
Propellant:
Type........................................ IMR 7013 or WC 875
Weight........................................ 500 gr

**Performance:**

Chamber pressure........................ 51,000 psi
Velocity........................................ 2680 fps, 78 ft from muzzle

**Shipping and Storage Data:**

Quantity-distance class/
SCG........................................ 1.4C
Storage code................................ Class V
DOT shipping class...................... C
DOT designation.......................... SMALL ARMS AMMUNITION
Drawing number.......................... 7258849

**References:**

SB 700-20
TM 9-1300-206

THIS PAGE INTENTIONALLY LEFT BLANK

## CARTRIDGE, 20MM, TARGET PRACTICE, M204

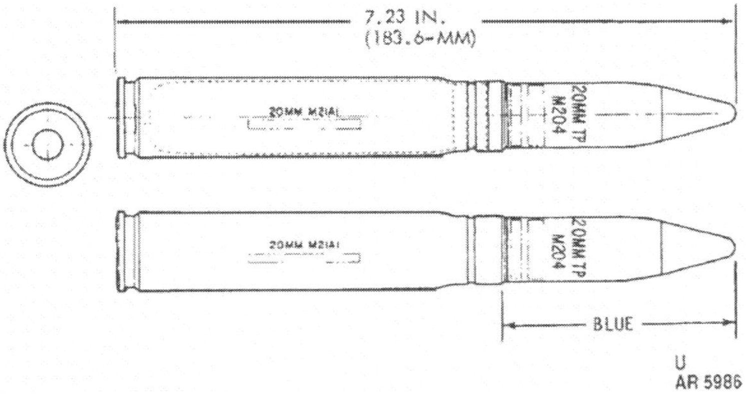


**Type Classification:**

STD - AMCTC 4762.

**Use:**

Guns, 20mm, M3 and M24A1. The cartridge with inert projectile is used in training, gun testing, and gun system function testing.

Description:

TARGET PRACTICE Cartridge. The cartridge has a blue painted projectile. The body has a cavity to control the weight and the aluminum nose is swaged to the body.

**Function:**

No specific function takes place upon impact except target penetration.

**Tabulated Data:**

| | |
|---|---|
| DODAC | 1305-A777 |
| UNO serial number | 0339 |
| UNO proper shipping name | Cartridges for weapons, inert projectiles |
| Weight | 4000 gr |
| Length (max) | 7.23 in. (183.6 mm) |
| Tracer | NA |
| Primer | Percussion, M36A2 |
| Fuze | NA |
| Explosive: | |
| Type | NA |
| Weight | NA |
| Incendiary: | |
| Type | NA |
| Weight | NA |
| Propellant: | |
| Type | WC 875 or IMR 7013 |
| Weight | MBR |

**Performance:**

| | |
|---|---|
| Chamber pressure (max) | 51,000 (copper) |
| Velocity | 2680 fps, 78 ft from muzzle |

**Shipping and Storage Data:**

| | |
|---|---|
| Quantity-distance class/ SCG | 1.4C |
| Storage | Class V |

DOT shipping class ...................... C
DOT designation .......................... SMALL ARMS AMMUNITION
Drawing number........................... 7259106

**References:**

SB 700-20
TM 9-1300-206

## CARTRIDGE, 20MM, TARGET PRACTICE -TRACER, M206A1

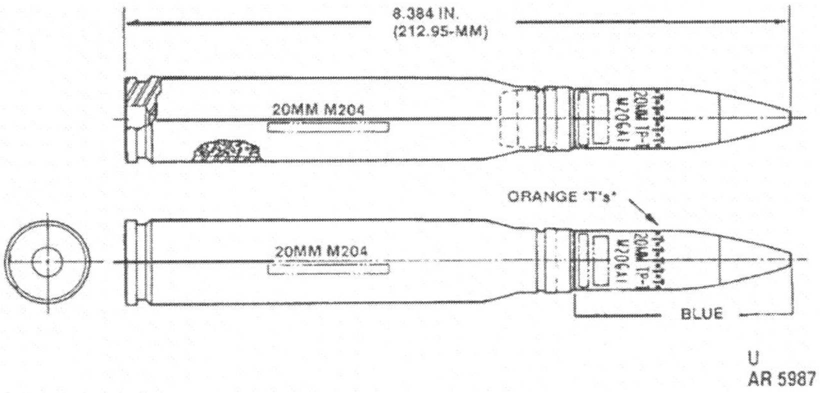


**Type Classification:**

STD - AMCTC 4762.

**Use:**

Gun, 20mm, Automatic, M139. The cartridge is used in training and operation of gun systems with tracer characteristics for visual sighting to target.

**Description:**

TARGET PRACTICE-TRACER Cartridge. The cartridge has a blue painted projectile body with a row of orange "T's" at the front of the projectile body.

**Tabulated Data:**

| | |
|---|---|
| DODAC | 1305-A790 |
| UNO serial number | 0339 |
| UNO proper shipping name | Cartridges for weapons, inert projectiles |
| Weight | 4886 gr |
| Length | 8.384 in. (212.95 mm) |
| Tracer | R284 |
| Primer | Percussion, MI15 |
| Fuze | NA |
| Explosive: | |
| Type | NA |
| Weight | NA |
| Incendiary: | |
| Type | NA |
| Weight | NA |
| Propellant: | |
| Type | CR 8325 |
| Weight | MBR (585-605 gr) |

**Performance:**

| | |
|---|---|
| Chamber pressure (max) | 54,000 (copper) |
| Velocity | 3430 fps, 78 ft from muzzle |

**Shipping and Storage Data:**

| | |
|---|---|
| Quantity-distance class/ SCG | 1.4C |
| Storage code | Class V |
| DOT shipping class | C |
| DOT designation | SMALL ARMS AMMUNITION |
| Drawing number | 10542655 |

**References:**

SB 700-20
TM 9-1300-206

THIS PAGE INTENTIONALLY LEFT BLANK

## CARTRIDGE, 20MM, HIGH EXPLOSIVE INCENDIARY, M210

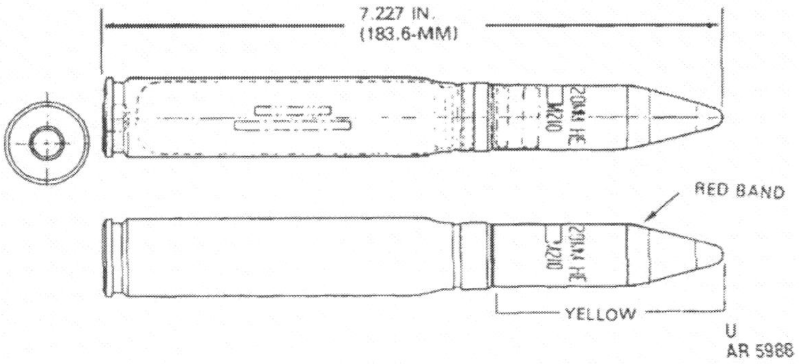


**Type Classification:**

STD - AMCTC 5433.

**Use:**

Guns, 20mm, M3 and M24A1. The projectile is effective against aircraft and other targets that are not heavily armored.

**Description:**

HIGH EXPLOSIVE INCENDIARY Cartridge. The cartridge has a yellow painted projectile body with a red band, approximately ¼ inch wide, at forward end. The projectile has a steel body with a cavity filled with RDX and aluminum and a point-detonating (PD) bore-safe fuze.

**Function:**

The projectile fuze arms in flight. Upon impact, the fuze detonates the projectile causing fragmentation of the body with an incendiary effect.

**Tabulated Data:**

DODAC ........................................ 1305-A863
UNO serial number ...................... 0009
UNO proper shipping
name ........................................ Ammunition, incendiary

Weight (approx) ........................... 4000 gr
Length (max)................................ 7.227 in. (183.6 mm)
Tracer .......................................... NA
Primer.......................................... Percussion, M36A2
Fuze............................................. PD, M505A3
Explosive:
Type ......................................... H761
Weight ...................................... 120 gr
Incendiary:
Type ......................................... NA
Weight ...................................... NA
Propellant:
Type ......................................... WC 875 or IMR 7013
Weight ...................................... MBR

**Performance:**

Chamber pressure (max).............. 51,000 psi (copper)
Velocity........................................ 2680 ± 50 fps, 78 ft from muzzle

**Shipping and Storage Data:**

Quantity-distance class/
SCG.......................................... 1.2E (0.4)
Storage code................................ Class V
DOT shipping class ...................... A

| | |
|---|---|
| DOT designation | AMMUNITION FOR SMALL ARMS WITH EXPLOSIVE PROJECTILES |
| Drawing number | 7259105 |

**References:**

SB 700-20
TM 9-1300-206

## CARTRIDGE, 20MM, TARGET PRACTICE-TRACER, M220

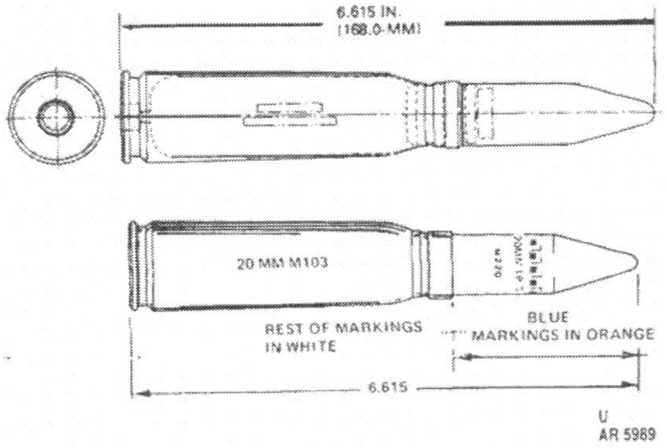


**Type Classification:**

STD - AMCTC 7404.

**Use:**

Guns, 20mm, M39, M61 and M168. The cartridge is for practice and training.

**Description:**

TARGET PRACTICE-TRACER Cartridge. The cartridge is ballistically and physically identical to the M55A2 cartridge except that a tracer element has been added to the base of the projectile. The projectile is painted blue with orange colored markings.

**Function:**

The projectile is similar to the M56A3 projectile, but has no high explosive incendiary charge and is assembled with a dummy fuze. The projectile does contain a tracer composition which burns visibly for a minimum of 1.9 seconds.

**Tabulated Data:**

DODAC.......................................1305-A651
UNO serial number ......................0339
UNO proper shipping
name..........................................Cartridges for weapons, inert projectiles
Weight .........................................3935 gr
Length.........................................6.615 in. (168 mm)
Tracer .........................................R505 or R403
Primer..........................................Electric
Fuze.............................................NA
Explosive:
Type .........................................NA
Weight ......................................NA
Incendiary:
Type .........................................NA
Weight ......................................NA
Propellant:
Type .........................................WC 870 or WC 872
Weight ......................................MBR (585-605 gr)

**Performance:**

Chamber pressure........................61,500 psi
Velocity........................................3380 fps, 78 ft from muzzle

**Shipping and Storage Data:**

Quantity-distance class/
SCG.........................................1.4C
Storage code................................Class V
DOT shipping class ......................A
DOT designation ..........................SMALL ARMS AMMUNITION
Drawing number...........................7259146

**References:**

SB 700-20
TM 9-1005-286-10
TM 9-1300-206

THIS PAGE INTENTIONALLY LEFT BLANK

## CARTRIDGE, 20MM, HIGH EXPLOSIVE INCENDIARY-TRACER, M242/M242A1

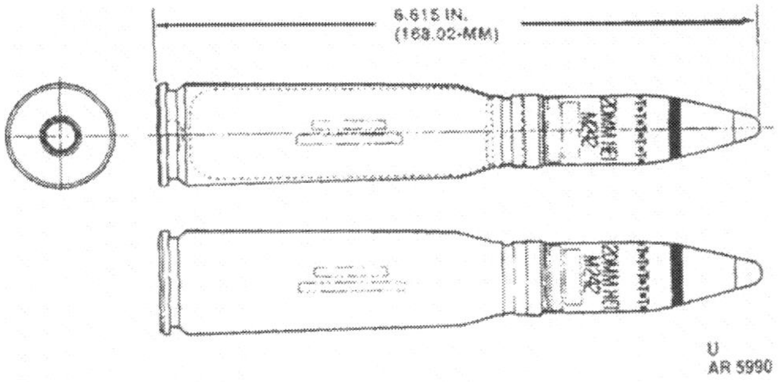


### Type Classification:

M242: STD - AMCTC 6445.
M242A1: No Army requirement.

### Use:

Guns, 20mm, M39 and M61.

### Description:

HIGH EXPLOSIVE INCENDIARY-TRACER Cartridge. The cartridge has the same basic structure as the M56A3 cartridge with the addition of a tracer element.

### Function:

The projectile has the same function as the M56A3 high explosive incendiary (HEI) cartridge projectile. The tracer is ignited by the propellant when the round is fired.

### Difference Between Models:

The models differ in the method of loading the projectiles. The M242 has the HE mix and the incendiary mix combined in one pellet with the tracer mix added at the base of the projectile. The M242A1 projectile has the incendiary loaded as a separate pellet and the HE mix is also loaded as a separate pellet, then the tracer pellet is added. This improves fire start capability.

### Tabulated Data:

| | |
|---|---|
| DODAC | 1305-A659 |
| UNO serial number | 0321 |
| UNO proper shipping name | Cartridges for weapons, with bursting charge |
| Weight | 3865 gr |
| Length | 6.615 in. (168.02 mm) |
| Tracer | R505 |
| Primer | Electric, M52A3B1 |
| Fuze | PD, M505A3 |
| Explosive: | |
| Type | H7G1 |
| Weight | 1455 gr |
| Incendiary: | |
| Type | I136 |
| Weight | 20 gr |
| Propellant: | |
| Type | WC 870 or WC 872 |
| Weight | MBR (585 or 605 gr) |

### Performance:

| | |
|---|---|
| Chamber pressure | 60,500 psi |
| Velocity | 3380 fps, 78 ft from muzzle |

### Shipping and Storage Data:

| | |
|---|---|
| Quantity-distance class/ SCG | 1.2E (0.4) |
| Storage code | Class V |
| DOT shipping class | A |

| | |
|---|---|
| DOT designation | SMALL ARMS AMMUNITION WITH EXPLOSIVE PROJECTILES |
| Drawing number | 7259268 (M242); 9341740 (M242A1) |

**References:**

SB 700-20
TM 9-1300-206

## CARTRIDGE, 20MM, HIGH EXPOSIVE INCENDIARY-TRACER, SELF-DESTROYING, M246/M246A1

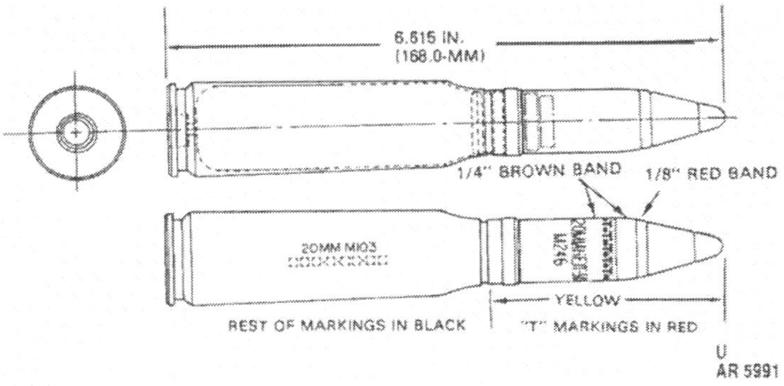


**Type Classification:**

M246: STD - LCC-B MSR 03816015.
M246A1: STD - LCC-A MSR 03816015.

**Use:**

Gun, 20mm, M168. The cartridge is for use against aerial targets.

**Description:**

HIGH EXPLOSIVE INCENDIARY-TRACER, SELF-DESTROYING Cartridge.

**Function:**

The cartridge is assembled with an M505A3 point-detonating (PD) fuze. The projectile, in addition to the M505A3 fuze, consists of a high explosive incendiary (HEI) charge, self-destroying (SD) relay charge, and a tracer charge. These charges form the SD chain. The tracer burns for about 3 to 7 seconds, whereupon the relay charge is ignited which detonates the HEI charge low order. If impact with target occurs prior to self-destruction, the M505A3 fuze causes the HEI charge to detonate high order.

**Difference Between Models:**

The basic difference between the models is in the loading of the projectiles. In the M246 projectile, the incendiary mix and the HE mix are combined in one pellet. In the M246A1 projectile, the incendiary is loaded as a separate pellet. The HE mix is also loaded as a separate pellet along with the SD relay charge pellet and the tracer charge pellet. This improves fire start capability.

**Tabulated Data:**

| | |
|---|---|
| DODAC: | |
| M246 | 1305-A791 |
| M246A1 | 1305-A792 |
| UNO serial number | 0321 |
| UNO proper shipping name | Cartridges for weapons, with bursting charge |
| Weight | 4070 gr |
| Length | 6.615 in. (168 mm) |
| Tracer | R505 |
| Primer | Electric, M52A3B1 |
| Fuze | PD, M505A3 |
| Explosive: | |
| Type | H761 |
| Weight | 115 gr |
| Incendiary: | |
| Type | NA |
| Weight | NA |
| Propellant: | |
| Type | WC 870 |
| Weight | MBR (585-605 gr) |

**Performance:**

Chamber pressure........................ 60,500 psi
Velocity........................................ 3380 fps, 50 ft from muzzle

**Shipping and Storage Data:**

Quantity-distance class/ SCG.......................................... 1.2E (0.4)
Storage code................................ Class V
DOT shipping class...................... A
DOT designation.......................... AMMUNITION FOR SMALL EXPLOSIVE PROJECTILES
Drawing number:
M246 ........................................ 10542384
M246A1 .................................... 9333764

**References:**

SB 700-20
TM 9-1005-286-10
TM 9-1300-206

## CARTRIDGE, DUMMY, 20MM, XM254

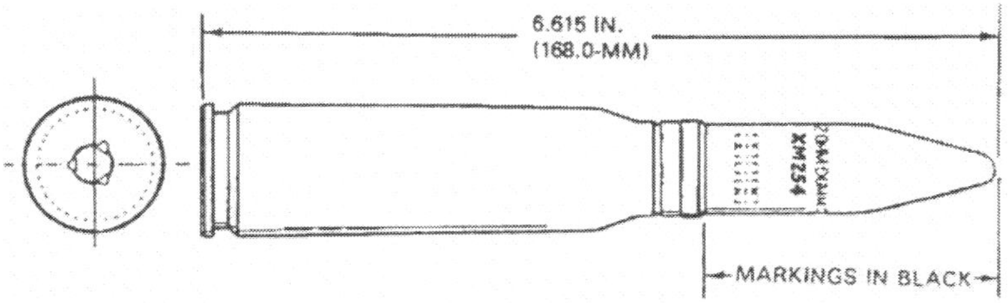


U
AR 5992

**Type Classification:**

Not applicable.

**Use:**

Guns, 20mm, M39, M61, M139, M168, M195 and XM197. The cartridge is for training of crews. The plastic material, instead of steel, reduces wear on gull components and fire mechanisms.

**Description:**

DUMMY Cartridge. The cartridge is of plastic construction.

**Tabulated Data:**

DODAC........................................ 1305-A895
Weight ......................................... 3850 gr
Length.......................................... 6.615 in. (168 mm)
Tracer ........... .............................. NA
Primer.......................................... NA
Fuze............................................. NA
Explosive:
  Type ......................................... NA
  Weight ...................................... NA
Incendiary:
  Type ......................................... NA
  Weight ...................................... NA
Propellant:
  Type ......................................... NA
  Weight ...................................... NA

**Performance:**

Chamber pressure........................ NA
Velocity ....................................... NA

**Shipping and Storage Data:**

Quantity-distance class/
  SCG.......................................... NA
Storage code................................ NA
DOT shipping class ...................... NA
DOT designation .......................... NA
Drawing number........................... 7259254

**References:**

SB 700-20
TM 9-1005-246-24
TM 9-1005-286-10
TM 9-1300-206

THIS PAGE INTENTIONALLY LEFT BLANK

## CARTRIDGE, 20MM, HIGH EXPLOSIVE INCENDIARY-TRACER, M599 FUZE POINT-DETONATING SELF-DESTROYING M594

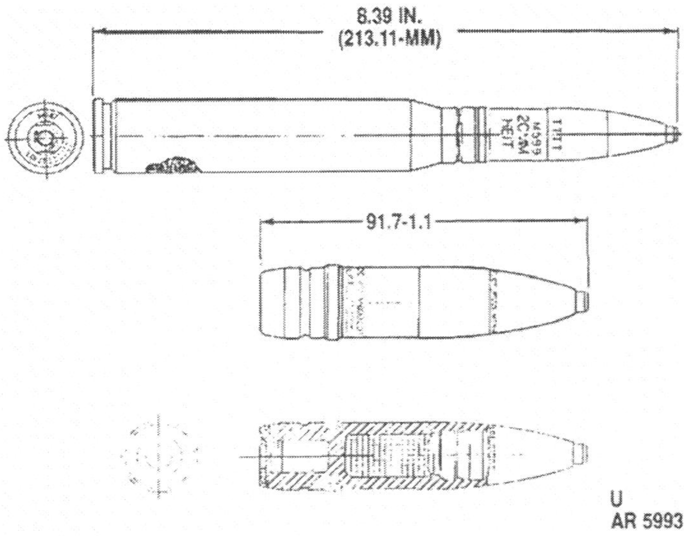


**Type Classification:**

STD - AMCTC 6391.

**Use:**

Gun, M139.

**Description:**

HIGH EXPLOSIVE INCENDIARY-TRACER Cartridge. The cartridge has a point-detonating (PD) self-destroying (SD) fuze M594. The projectile is a hollow body filled with 120 grams of high explosive incendiary (HEI) mixture with a tracer tablet at the base and the PD SD fuze screwed onto the nose.

**Function:**

When fired, the weapon firing pin initiates the primer which ignites the propelling charge. Gases produced by the burning propellant propel the projectile. The projectile shall function with high order detonation upon impact or, if the projectile does not impact on intended target within 4 to 7 seconds, a SD mechanism will cause high order detonation of the projectile.

**Tabulated Data:**

| | |
|---|---|
| DODAC | 1305-A787 |
| UNO serial number | 0321 |
| UNO proper shipping name | Cartridges for weapons, with bursting charge |
| Type | HEI-T |
| Weight (approx) | 317 g |
| Length | 8.39 in. (213.11 mm) |
| Projectile (steel): | |
| Tracer | Tablet, 1.17 g |
| Explosive (HEI mix) | 120 g |
| Fuze | PD SD, M594, aluminum |
| Color | Yellow with red "T" band, black markings |
| Cartridge case | Steel, M187 |
| Propellant | VTL-1375-502-52 g |
| Primer percussion | M103 |

**Performance:**

Chamber pressure........................ 49,500 psi
Velocity........................................ 3460 fps, 78 ft from muzzle

**Shipping and Storage Data:**

Quantity-distance class/
SCG.......................................... 1.2E
Storage code................................ Class IV
DOT shipping class...................... A
DOT designation.......................... SMALL ARMS AMMUNITION WITH EXPLOSIVE PROJECTILE

M528 shipping and storage container:
Material.................................... Steel
Dimensions............................... 14-9/16 x 9-39/64 x 4-5/16 in.
Cube........................................ 0.35 cu ft
Drawing number C/R................ 10535815

**Packing:**

25 cartridges, 20mm linked with M601 armor piercing incendiary-tracer, cartridges in ratio with 3 M599 to 2 M601 in shipping and storage containers M528. Two shipping and storage containers (50 cartridges) in a wire-bound box.

**References:**

DARCOM 700-3-2
TM 9-1300-206

## CARTRIDGE, 20MM, ARMOR PIERCING INCENDIARY-TRACER, M601

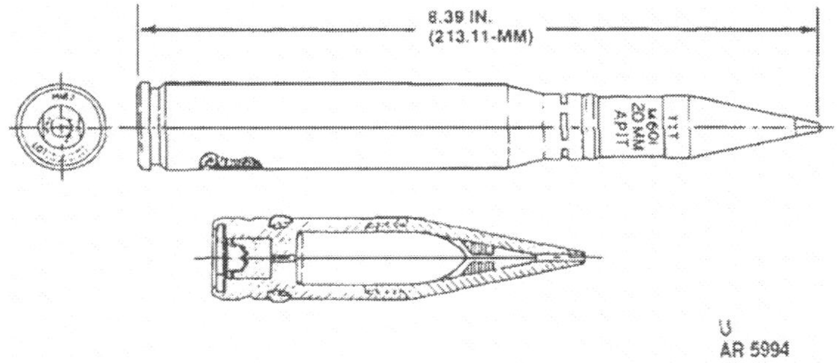


**Type Classification:**

STD - AMCTC 4762.

**Use:**

Gun, M139.

**Description:**

ARMOR PIERCING INCENDIARY-TRACER Cartridge. The tungsten-carbide core is the main part of the projectile, and is the penetrator. The ballistic cap (cgive) contains an incendiary pellet. A tracer tablet is at the base of the projectile. The projectile body is painted black with white markings and with red painted "T's" encircle the ogive (signifying tracer).

**Function:**

When fired, the weapon firing pin initiates the primer which ignites the propelling charge. Gases produced by the burning propellant propel the projectile. Upon impact with the target, the nose (ogive) ruptures and the incendiary composition ignites on exposure to the air.

**Tabulated Data:**

| | |
|---|---|
| Complete round: | |
| UNO serial number | 0009 |
| UNO proper shipping name | Ammunition, incendiary |
| DODAC | 1305-A786 |
| Type | API-T |
| Weight | 310 g |
| Length | 8.39 in. (213.112 mm) |
| Cannon used with | M139 |
| Projectile (111 g): | |
| Penetrator (core) | Tungsten-carbide |
| Body | Aluminum alloy, 22.5 g |
| Windshield | Aluminum, 9.4 g |
| Incendiary | XM71 pellet, 0.45 g |
| Tracer | Tablet, 1.17 g |
| Fuze | NA |
| Color | Black with white marking with "T" red bound |
| Cartridge case | Steel, M187 |
| Propellant (VTL 1375-502): | |
| Weight (approx) | 53 g |
| Primer (percussion) | M103 |

**Performance:**

Chamber pressure........................ 49,500 psi
Velocity........................................ 3610 fps, 78 ft from muzzle

**Shipping and Storage Data:**

Quantity-distance class/ SCG.......................................... 1.4G
Storage code................................ Class IV
DOT shipping class...................... A
DOT designation.......................... SMALL ARMS AMMUNITION WITH INCENDIARY PROJECTILE

M528 shipping and storage container:
Material.................................... Steel
Dimensions............................... 14-9/16 x 9-39/64 x 4-5/16 in.
Cube......................................... 0.35 cu ft
Drawing number C/R................. 10535805

**Packing:**

25 cartridges, 20mm linked with M599 high explosive incendiary-tracer cartridges in a ratio of 3 M599 to 2 M601 in shipping and storage container M528. Two containers in a wrap-around crate.

**References:**

DARCOM 700-3-2
TM 9-1300-206

## CARTRIDGE, 20MM, MULTIPURPOSE TRACE SELF-DESTROYING, M940

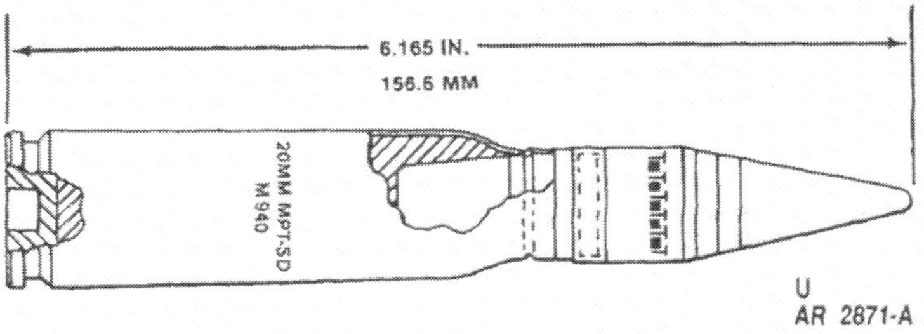


**Type Classification:**

TC STD A - Approved March 13, 1989.

**Use:**

The existing Vulcan Air Defense System (VADS) and Product Improved Vulcan Air Defense System (PIVADS) M246 ammunition is incapable of providing defense against the modern enemy threat at combat ranges greater than 1,600 meters because of inherent performance limitations including poor flight characteristics, self-destroying ranges, and terminal effectiveness. Therefore, the Army decided to pursue acquiring a new round of ammunition for VADS/PIVADS which would exhibit better exterior ballistics and greater lethality than the existing M246 High Explosive Incendiary Tracer Self-Destroying (HEI-T-SD) ammunition.

**Description:**

MULTIPURPOSE TRACE SELF-DESTROYING Cartridge. The cartridge is an improved fuzeless pyrotechnic initiated 20mm HE cartridge. The projectile consists of a heavy walled steel body charged with explosive mix and an incendiary mix, and an aluminum nose charged with another incendiary mix. The projectile also has safe and arming (S&A), tracer, and self-destroying which is initiated by heat from combustion of the tracer composition. The lack of fuze permits an enhanced aerodynamic ogive shape which results in a reduction in drag factor and reduced time of flight, thereby extending the effectiveness of the VADS from 1,200 meters to 2,000 meters against aircraft targets. The M940 will replace the 20mm M246 HEI-T-SD as the primary air defense cartridge for the VADS and any of its improved versions.

**Function:**

The projectile consists of a steel body charged with a HE mix, Composition A-4 (MILC-440), and an incendiary mix, Type I (equivalent I-68). The projectile has an aluminum nose charged with Type II incendiary (equivalent I-136). The hardened forebody of the projectile provides for defeat of lightweight armor (3/8 RHA at well over 500 m). The incendiaries provide a built-in delay when initiated on the thin skin of an aircraft, and the projectile explodes approximately 1 foot behind the skin. The M94 cartridge, therefore, is primarily intended for air defense but has a secondary roll against light armored vehicles. The M940 contains a chemical S&A intended to reduce the probability of functioning on a 0.04 aluminum plate at 1 meter from the muzzle. When initiated, all projectile fragments go forward.

**Tabulated Data:**

Complete round:

| | |
|---|---|
| UNO serial number | 0321 |
| UNO proper shipping name | Cartridges for weapons, with bursting charge |
| DODAC | 1305-A710 |
| Type | MPT-SD |
| Weight | 264 g |
| Length (max) | 6.165 in. (156.6 mm) |
| Cannon used with | M168 |

Projectile:
- Tracer-modified ........................ R505
- Primer-electric .......................... M52A3B1
- Fuze ......................................... NA

Explosive:
- Type ....................................... Comp A-4
- Weight...................................... 4g

Incendiary:
- Type ......................................... I
- Weight...................................... 1.2 g

Incendiary:
- Type ........................................ II
- Weight ..................................... 3 g

Incendiary:
- Type ........................................ Potassium Perchlorate
- Weight...................................... 0.2 g

Propellant:
- Type ......................................... WC 866
- Weight ..................................... 37 g

**Performance:**

Chamber pressure (max) ................ 60,500 psi
Velocity .......................................... 3350 ± 45 fps, 78 ft from muzzle

**Shipping and Storage Data:**

DOD hazard class/division/
SCG ....................................... 08 (1.2E)
DOT shipping class ......................... Class A explosive
DOT designation ............................. CARTRIDGE FOR WEAPONS
Drawing number .............................. 12551767
NSN ........................................... 1305-01-283-2134

**References:**

TM 9-1300-206

# CHAPTER 15

# 25MM CARTRIDGES

THIS PAGE INTENTIONALLY LEFT BLANK

## CARTRIDGE, 25MM, ARMOR PIERCING DISCARDING SABOT-TRACER, M791

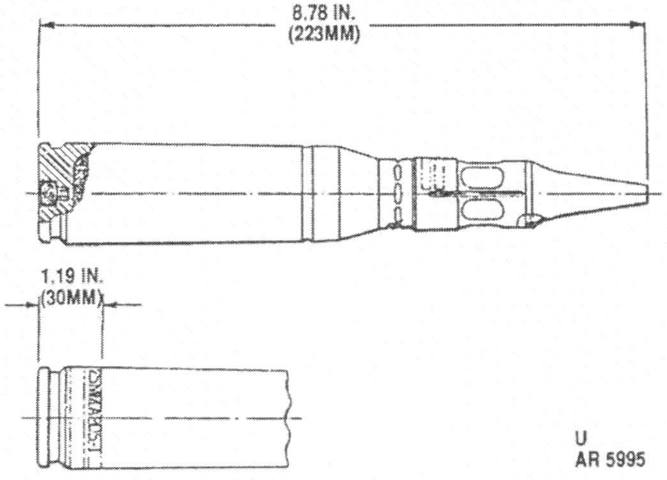


### Type Classification:

STD - MSR 12796010.

### Use:

Automatic cannon, 25mm, M242.

### Description:

ARMOR PIERCING DISCARDING SABOT-TRACER Cartridge. The cartridge is a fixed type, percussion primed 25mm ammunition, intended as the armor piercing service round to successfully engage lightly armored vehicles. The cartridge consists of a sabot encapsulated projectile body, crimped to a steel cartridge case containing 98 grams of extruded propellant, M115 percussion primer and booster pellet. The projectile body consists of a solid tungsten alloy penetrator, pressed on aluminum windshield, pressed-in tracer pellets, molded discarding-type nylon sabot, staked aluminum base and pressed-on polythylene nose cap.

### Function:

When fired, the weapon firing pin initiates the primer which ignites the propelling charge. Gases produced by the burning propellant propel the projectile from the cannon and ignite the tracer. Setback, centrifugal and air pressure forces cause the sabot to discard upon leaving the gun tube. The penetrator (core) is spin stabilized and penetrates the target solely by kinetic energy.

### Tabulated Data:

| | |
|---|---|
| Complete round: | |
| DODAC | 1305-A974 |
| UNO serial number | 0339 |
| UNO proper shipping name | Cartridges for weapons, inert projectile |
| Type | APDS-T |
| Weight | 458 ± 15 g |
| Length | 8.78 in. (223 mm) |
| Cannon used with | M242 |
| Projectile: | |
| Penetrator (core) | Tungsten-alloy |
| Sabot | Nylon, aluminum base |
| Nose cap | Polyethylene |
| Tracer | NA |
| Explosive | NA |
| Fuze | NA |
| Color | Black with white markings |
| Cartridge case | C1030 steel |
| Propellant: | |
| Type | Radford AP225 |
| Weight | 0.221 lb |
| Primer percussion | M115 |

### Performance:

| | |
|---|---|
| Chamber pressure | 56,000 psi |
| Range | 15,000 m |
| Muzzle velocity | 1330 mps, 4370 fps |

**Shipping and Storage Data:**

Quantity-distance class/
SCG ........................................ 1.4C
Storage code .................................. Class IV
DOT shipping class ........................ A
DOT designation ............................ AMMUNITION FOR CANNON WITH SOLID PROJECTILE
M621 shipping and storage:
Material..................................... Polypropelene plastic
Weight (tactically packed) ............................. 47.7 lb
Weight (empty) ........................ 7.94 lb
Dimensions .............................. 13.8 x 13.8 x 5.5 in.
Drawing number C/R................. 12013533

**Packing:**

Tactical packing unit 30 linked cartridge in two 15-round belts packed nose-to-nose with two plastic separators in plastic shipping and storage container M621.

**References:**

DARCOM 700-3-2
TM 9-1300-206

## CARTRIDGE, 25MM, HIGH EXPLOSIVE INCENDIARY-TRACER, M792 WITH FUZE POINT-DETONATING SELF-DESTROYING M758

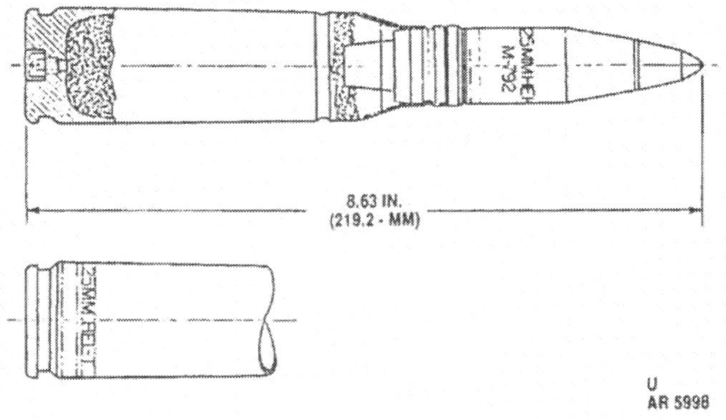


**Type Classification:**

STD - MSR 12796010.

**Use:**

Automatic cannon, 25mm, M242.

**Description:**

HIGH EXPLOSIVE INCENDIARY-TRACER Cartridge. The cartridge is a fixed type, percussion primed 25mm munition, intended as the high explosive service round to successfully engage unarmored ground targets. The cartridge consists of a high explosive incendiary (HEI) filled, one piece projectile body that is crimped to a steel cartridge case containing 90 grams of ball propellant, M115 percussion primer and booster pellet. The projectile body consists of a hollow C1045 steel body, screwed on M758 fuze, swaged iron rotating band, 32 grams of H761 HEI mix and pressed-in tracer pellets. Projectile length, including tracer boom, varies by manufacturer from 3.94 to 4.66 inches.

**Function:**

When fired, the primer is mechanically initiated by the weapon firing pin. The primer output ignites the booster pellet which, in turn, ignites the propellant charge. The projectile is launched at a velocity of 1100 meters per second, achieving a maximum effective range of 3000 meters. Upon impact, the fuze functions and the HEI filler detonates dispersing the incendiary mixture. The self-destroying requirement is 6.2 seconds per minute and prior to impact.

**Tabulated Data:**

| | |
|---|---|
| Complete round: | |
| DODAC | 1305-A975 |
| UNO serial number | 0321 |
| UNO proper shipping name | Cartridges for weapons, with bursting charge |
| Type | HEI-T |
| Weight | 1.11 lb |
| Length | 8.63 in. (219.2 mm) |
| Projectile: | |
| Body | Steel, C1045 |
| Tracer | Pellet |
| Explosive (HEI mix) | 4761, 32 g |
| Fuze | PD SD, M758 |
| Color | Yellow with red band and black markings |
| Cartridge case | Steel, C1030 |
| Propellant: | |
| Type | WC 890 |
| Weight | 90 g |
| Primer percussion | M115 |

**Performance:**

Range ........................................... 3000 m
Muzzle velocity ............................. 1100 mps, 3610 fps
Self-destruct .................................. 6.2 mps

**Shipping and Storage Data:**

Quantity-distance class/
SCG ........................................... 1.2E
Storage code .................................. Class IV
DOT shipping class ........................ A
DOT designation ............................ AMMUNITION FOR CANNON WITH EXPLOSIVE PROJECTILE

M621 shipping and storage container:
Material.................................... Polypropelene
Weight (tactically packed) ................................... 47.7 lb
Weight (empty) ........................ 7.94 lb
Dimensions .............................. 13.8 x 13.8 x 5.5 in.
Drawing number........................ 12013230

**Packing:**

Tactical packing unit 30 linked cartridge in two 15-round belts packed nose-to-nose with two plastic separators in plastic shipping and storage container M621.

**References:**

DARCOM 700-3-2
TM 9-1300-206

## CARTRIDGE, 25MM, TARGET PRACTICE-TRACER, M793

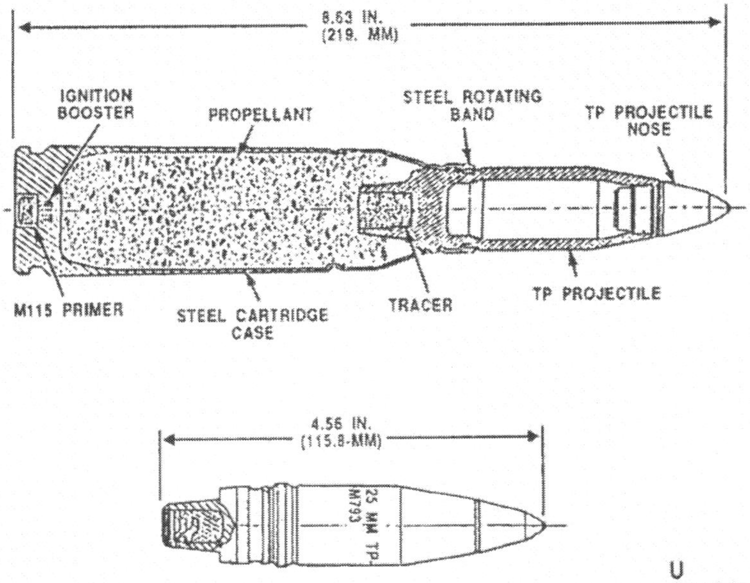


U
AR 5999

**Type Classification:**

STD - MSR 12796010.

**Use:**

Automatic cannon, 25mm, M242.

**Description:**

TARGET PRACTICE-TRACER Cartridge. The cartridge is a fixed type, percussion primed 25mm munition, intended as a substitute for the high explosive incendiary with tracer service round for target practice in training, gun testing, and gun system functioning. The cartridge consists of a hollow, two piece projectile body that is crimped to a steel cartridge case containing 90 grams of ball propellant, M115 percussion primer and booster pellet. The projectile body consists of a hollow C1035 steel body screw-on steel nose piece, swaged iron rotating even band and pressed-in tracer pellets.

**Function:**

The cartridge functions the same as the M792, except upon impact the projectile is inert.

**Tabulated Data:**

Complete round:
- DODAC .................................. 1305-A976
- UNO serial number .................. 0339
- UNO proper shipping name ................................ Cartridges for weapons, inert projectile
- Type ....................................... TP-T
- Weight ..................................... 1.11 lb
- Length ................................... 8.63 in. (219 mm)

Projectile:
- Body ....................................... Steel, C1035/12/14
- Tracer ..................................... Pellet
- Explosive ................................. NA
- Fuze ........................................ NA
- Color ....................................... Blue with white markings
- Cartridge case .......................... C1030

Propellant:
- Type ....................................... WC 890
- Weight ..................................... 0.3 lb
- Primer percussion .................... M115

**Performance:**

Range ........................................... 6000 m, 196858E
Muzzle velocity ............................. 3609 fps, 78 ft from muzzle

**Shipping and Storage Data:**

Quantity-distance class/
SCG ......................................... 1.4C
Storage code .................................. 11 B
DOT shipping class ........................ Class B

DOT designation ............................ AMMUNITION FOR CANNON WITH EMPTY PROJECTILE

**Packing:**

Tactical packing unit 30 linked cartridge in two 15-round belts packed nose-to-nose with two plastic separators in plastic shipping and storage container M621.

**References:**

DARCOM 700-3-2
TM 9-1300-206

## CARTRIDGE, 25MM, DUMMY, M794

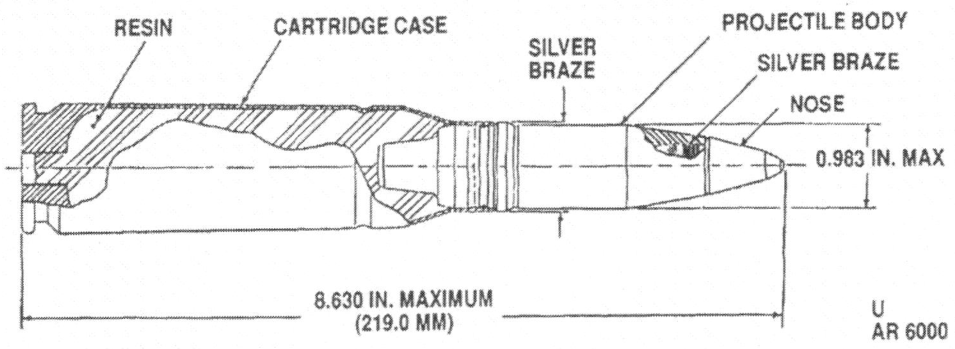


**Type Classification:**

STD - DA Letter December 1981.

**Use:**

Automatic cannon, 25mm, M242. The cartridge is used for non-firing system checkout for the 25mm automatic cannon M242 and training of the crew.

**Description:**

DUMMY Cartridge. The cartridge is a modified M793 target practice with tracer (TP-T) projectile silver brazed to a modified standard cartridge case and is completely inert. The cartridge case cavity is filled through an enlarged primer hole with an epoxy resin filler to approximate the weight of the M792 high explosive incendiary-tracer and M793 TP-T service cartridges. The threaded nose of the TP-T projectile is silver brazed to the projectile body to improve physical integrity. The external configuration of the M794 metal dummy cartridge (by use of standard components) is the same as that of the M792 HEI-T and M793 TP-T service cartridges.

**Tabulated Data:**

Complete round:
- DODAC .................................. 1305-A967
- Type ........................................ Dummy, inert
- Weight (w/o link) ....................... 1.11 lb
- Length .................................... 8.63 in. (219 mm)

Projectile:
- Body ........................................ Steel, C1035
- Tracer ...................................... NA
- Explosive ................................. NA
- Fuze ........................................ NA
- Color ....................................... Silver braze
- Cartridge case .......................... Steel, C1030

Propellant:
- Filler ........................................ Epoxy resin (polyurethane)
- Primer ...................................... NA

**Performance:**

Chamber pressure .......................... NA
Velocity .......................................... NA

**Shipping and Storage Data:**

Quantity-distance class/
- SCG ....................................... NA

Storage code .................................. NA
DOT shipping class .......................... NA
DOT designation .............................. NONEXPLOSIVE AMMUNITION

M621 shipping and storage container:
- Material .................................... Polypropelene plastic
- Length (max) ............................ 14.3 in.
- Width (max) .............................. 13.8 in.
- Height (max) ............................ 5.7 in.
- Packaged weight (30 linked rounds) .................... 47.7 lb

Wooden box:
- Length .................................... 32.5 in.
- Width ...................................... 11.5 in.
- Height ..................................... 12 in.
- Weight:
  - Packed (40 rounds with link) ....................... 85.3 lb
  - Packed (80 rounds without link) ................ 119 lb
- Drawing number ....................... 12013880

**Packing:**

Packing consists of 30 linked cartridges in two 15-round belts in M621 plastic shipping and storage containers, or Packing consists of 50 linked cartridges in a wooden shipping and storage container, or Packing consists of 80 tubed cartridges in a wooden shipping and storage container.

**References:**

DARCOM 700-3-2

## CARTRIDGE, 25.4MM, DECOY, M839

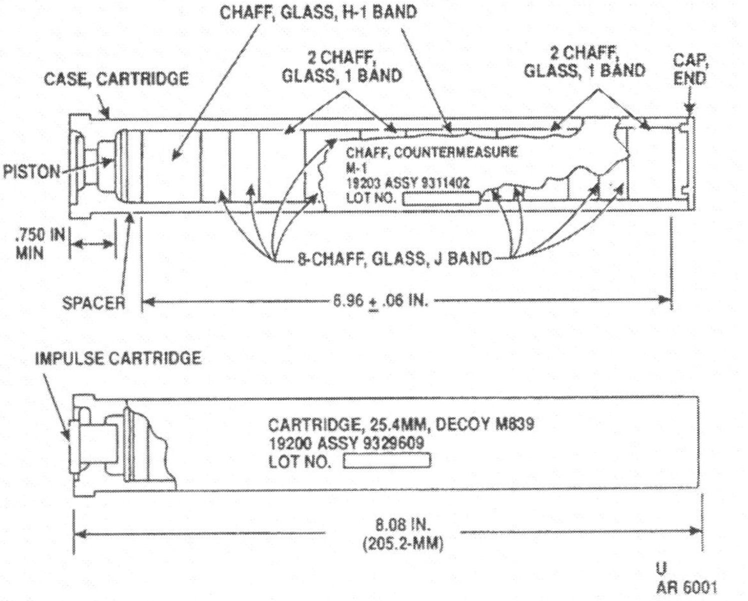


**Type Classification:**

STD - MSR 02816005.

**Use:**

The cartridge is a decoy to provide an effective survival counter-measure for Army aircraft against radar controlled weapon systems.

**Description:**

DECOY Cartridge. The cartridge is ammunition item consisting of a plastic outer case nominally 1-inch wide x 1-inch high x 8-inches long, a payload made up of approximately 390,000 strands of aluminum coated glass fibers. The payload is sectioned into several bundles of chaff dipoles. There are three different lengths of bundles in the payload. In addition, the item contains a plastic end cap, plastic piston, felt spacer, and impulse cartridge.

**Function:**

The cartridge is fired by electrical initiation of the impulse cartridge M796. The gases generated by the functioned impulse cartridge causes the piston to expel the chaff payload into the air where it blooms into a radar reflecting cloud that breaks "Lock On" by radar-controlled weapon systems.

**Tabulated Data:**

| | |
|---|---|
| Complete round: | |
| DODAC | 1305-A965 |
| UNO serial number | 0454 |
| UNO proper shipping name | Igniters |
| Type | Decoy |
| Weight | 0.34 lb |
| Length | 8.08 in. (205.2 mm) |
| Body | Plastic |
| Explosive | NA |
| Fuze | NA |
| Color | Unpainted |
| Drawing number | 9329609 |
| Specification number | MIL-C-63338(AR) |
| Payload: | |
| Chaff | Aluminum coated glass rods |
| Weight | 0.26 lb |
| Expelling charge: | |
| Type | Hercules bulls eye smokeless powder |
| Weight | 0.013 oz |
| Impulse cartridge | M796 |

**Shipping and Storage Data:**

Quantity-distance class/
SCG ........................................ 1.4S
DOT shipping class .................. C
DOT designation ...................... ELECTRIC SQUIB, CLASS C EXPLOSIVE, HANDLE CAREFULLY

Packing box:
Weight ...................................... 67 lb
Dimensions ............................... 14.25 x 13.27 x 12.8 in.

**Packing:**

Two PA19 metal containers containing 50 decoy cartridges each (width-size) in wood box.

**References:**

TM 9-1300-206

**CARTRIDGE, 25MM, TARGET PRACTICE DISCARDING SABOT-TARGET, M910**

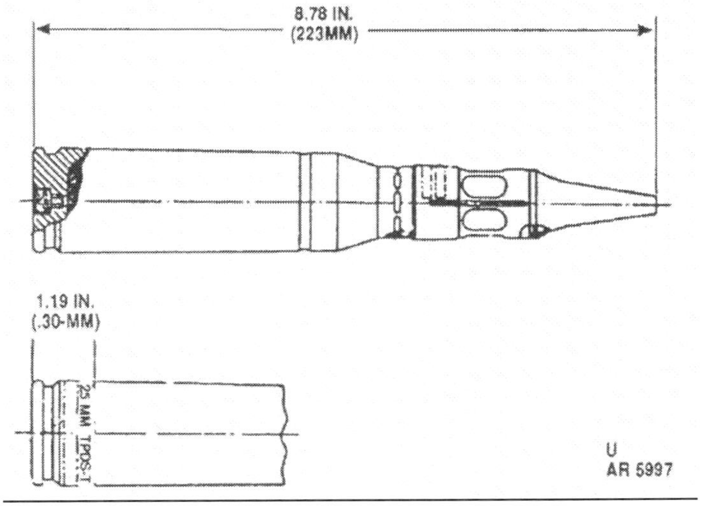


**Type Classification:**

To be approved.

**Use:**

Automatic cannon, 25mm, M242.

**Description:**

TARGET PRACTICE DISCARDING SABOT-TRACER Cartridge. The cartridge is a fixed type, percussion primed 25mm ammunition, intended as a training substitute for armor piercing service ammunition. The cartridge consists of a sabot encapsulated projectile body, crimped to a steel cartridge case containing 97 grams of extruded propellant, M115 percussion primer and booster pellet. The projectile body consists of a solid steel core, pressed-on aluminum windshield, pressed-in tracer pellets, molded discarding-type nylon sabot, staked aluminum base and pressed-on polyethylene nose cap.

**Function:**

When fired, the weapon firing pin initiates the primer which ignites the propelling charge. Gases produced by the burning propellant propel the projectile through the barrel and ignite the tracer. Centrifugal and air pressure forces cause the sabot, protective cap, and aluminum base to discard upon leaving the barrel. The high mass flow tracer decreases the base drag, allowing the spin-stabilized core to match the armor piercing round trajectory to a range of 2000 meters. Upon tracer burnout, the lightweight core loses velocity rapidly and eventually tumbles when its velocity goes below the speed of sound.

**Tabulated Data:**

| | |
|---|---|
| Complete round: | |
| DODAC | 1305-A940 |
| UNO serial number | 0328 |
| UNO proper shipping name | Cartridges for weapons, inert projectile |
| Type | TPDS-T |
| Weight | 421 ± 15 g |
| Length | 8.78 in. (223 mm) |
| Cannon used with | M242 |
| Projectile: | |
| Core | Steel |
| Sabot | Nylon, aluminum base |
| Nose cap | Polyethylene |
| Tracer | Magnesium |
| Explosive | NA |
| Fuze | NA |
| Color | Blue with white markings |
| Cartridge case | Steel, C1030 |
| Propellant: | |
| Type | Expro XPR51C4 |
| Weight | 0.214 lb |
| Primer Percussion | M115 |

**Performance:**

Chamber pressure .......................... 56,000 psi
Range 8000 m
Muzzle velocity .............................. 1545 mps, 5070 fps

**Shipping and Storage Data:**

DOD hazard class/SCG................... 1.2C (08)
Storage class ................................. Class B explosive
DOT designation ............................ AMMUNITION FOR CANNON WITH SOLID PROJECTILES
DOT label ....................................... Explosive B
UN serial number ............................ 0328
M621 shipping and storage:
Material........................................... Polypropylene plastic
Weight (tactically packed) .................................. 45 lb
Weight (empty) ............................. 8 lb
Dimensions ................................... 13.8 x 13.8 x 5.5 in.
Drawing number (for packing M910 in M621) ...................................... 12556515

**Packing:**

Tactical packing unit 30 linked cartridges in two 15-round belts packed nose-to-nose with two plastic separators in plastic shipping and storage container M621.

**References:**

DARCOM 700-3-2
TM 9-1300-206

## CARTRIDGE, 25MM, ARMOR PIERCING FIN STABILIZED DISCARDING SABOT-TRACER, M919

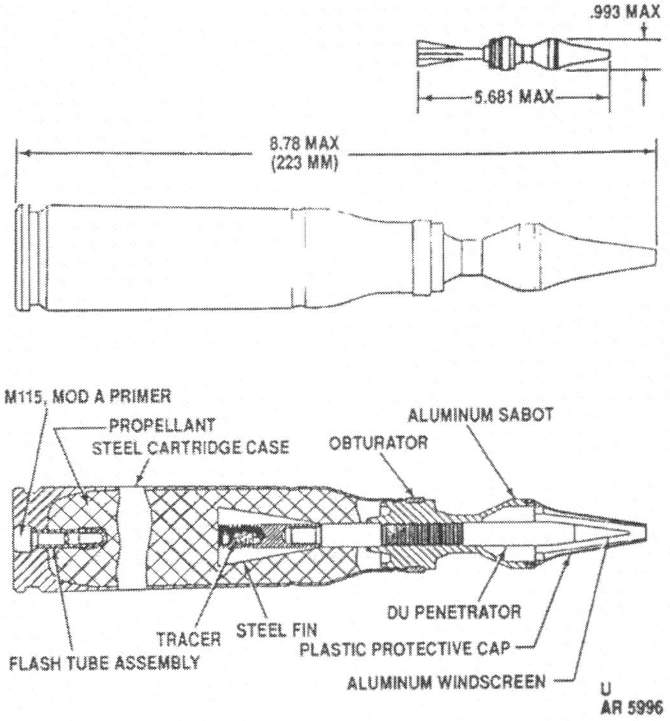


**Type Classification:**

STD.

**Use:**

Automatic cannon, 25mm, M242.

**Description:**

ARMOR PIERCING FIN STABILIZED DISCARDING SABOT-TRACER Cartridge. The cartridge is a fixed type, percussion primed 25mm ammunition, intended to supersede the M791 for use in combat. The cartridge consists of a subcaliber projectile inside an aluminum sabot, crimped to a steel cartridge case containing 97 grams of extruded propellant, modified M115 percussion primer and flash tube. The subprojectile consists of a depleted uranium penetrator with a pressed-on aluminum windscreen and a threaded steel fin with pressed-in tracer pellets. The three piece aluminum sabot utilizes a nylon obturator and a polyethylene nose cap.

**Function:**

When fired, the weapon firing pin initiates the primer which ignites the propelling charge. Gases produced by the burning propellant propel the projectile through the barrel and ignite the tracer. Centrifugal, air, and gas pressure forces cause the obturator, sabot, and nose cap to discard upon leaving the barrel. The low drag subcaliber projectile travels along its trajectory stabilized by the fin and a small amount of spin. The tracer allows visual contact with the fired round as it travels downrange. Upon impact, kinetic energy stored in the subprojectile causes it to punch through an armored target and inflict considerable damage inside, prohibiting any further use of the target.

**Tabulated Data:**

Complete round:
- DODAC .................................... 1305-A986
- UNO serial number .................. 0328
- UNO proper shipping name ................................ Cartridges for weapons, inert projectile
- Type ........................................ APFSDS-T
- Weight ...................................... 450 ± 15 g
- Length .................................... 8.78 in. (223 mm)
- Cannon used with .................... M242

Projectile:
- Core ......................................... Depleted uranium
- Sabot ....................................... Aluminum
- Nose cap ................................. Polyethylene
- Tracer ...................................... Magnesium
- Explosive .................................. NA
- Fuze ......................................... NA
- Color ........................................ Black and white obturator
- Cartridge case .......................... Steel, C1030

Propellant:
- Type ........................................ Hercules HES9053
- Weight ...................................... 98 ± 3 g
- Primer percussion .................... M115 Mod A

**Performance:**

Chamber pressure .......................... 56,000 psi
Muzzle velocity .............................. 1420 mps, 4659 fps

**Shipping and Storage Data:**

DOD hazard class/SCG.................. 1.2C
Storage code .................................. Class B explosive
DOT designation ............................ AMMUNITION FOR CANNON WITH SOLID PROJECTILES
UN serial number ............................ 0328 and 2912

**Packing:**

Tactical packing unit 30 linked cartridges in two 15-round belts packed nose-to-nose with two plastic separators in plastic shipping and storage container M621.

# CHAPTER 16

# 30MM CARTRIDGES

THIS PAGE INTENTIONALLY LEFT BLANK

## CARTRIDGE, 30MM, TARGET PRACTICE, M788

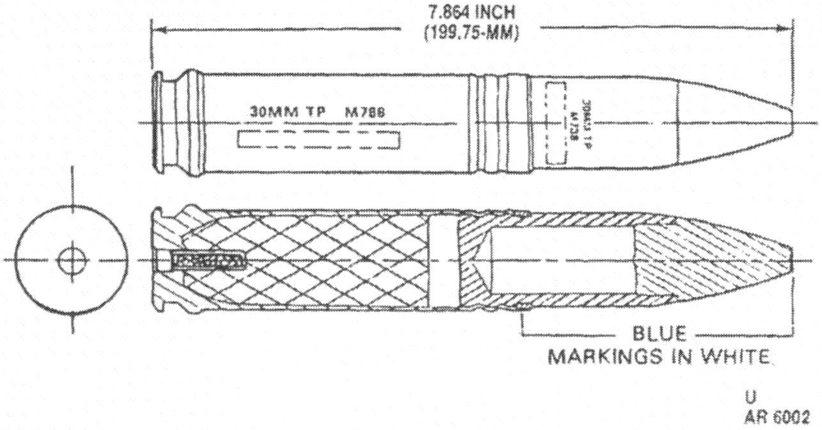


**Type Classification:**

STD.

**Use:**

Automatic cannon, 30mm, XM230 (U.S.), ADEN MK IV (NATO), DEFA 552/553 (NATO). The cartridge is used in target practice.

**Description:**

TARGET PRACTICE Cartridge. The cartridge has a blue painted projectile with white markings. The inert/solid projectile is a three piece assembly consisting of steel body with cavity, rotating band, and aluminum nose piece. The cartridge case is aluminum.

**Function:**

The cartridge consists of an inert, solid projectile without a fuze. No function takes place except penetration of target.

**Tabulated Data:**

DODAC ........ 1305-B118
UNO serial number ........ 0339
UNO proper shipping
name ........ Cartridges for weapons, inert projectile
Weight ........ 5371 gr
Length ........ 7.864 in. (199.75 mm)
Tracer ........ NA
Primer ........ Electric, PA520
Fuze ........ NA
Explosive:
Type ........ NA
Weight ........ NA
Incendiary:
Type ........ NA
Weight ........ NA
Propellant:
Type ........ WC 855
Weight ........ MBR, 50 gr

**Performance:**

Chamber pressure (avg)........ 40,600 to 44,950 psi (280-310 MPa)
Velocity ........ 2640 fps (805 mps)

**Shipping and Storage Data:**

Quantity-distance class/
SCG ........ 1.4C
Storage code ........ Class V
DOT shipping class ........ B
DO' designation ........ AMMUNITION FOR CANNON WITH SOLID PROJECTILES
Drawing number ........ 28114038

**References:**

SB 700-20
TM 9-1300-206

THIS PAGE INTENTIONALLY LEFT BLANK

## CARTRIDGE, 30MM, HIGH EXPLOSIVE DUAL PURPOSE, M789

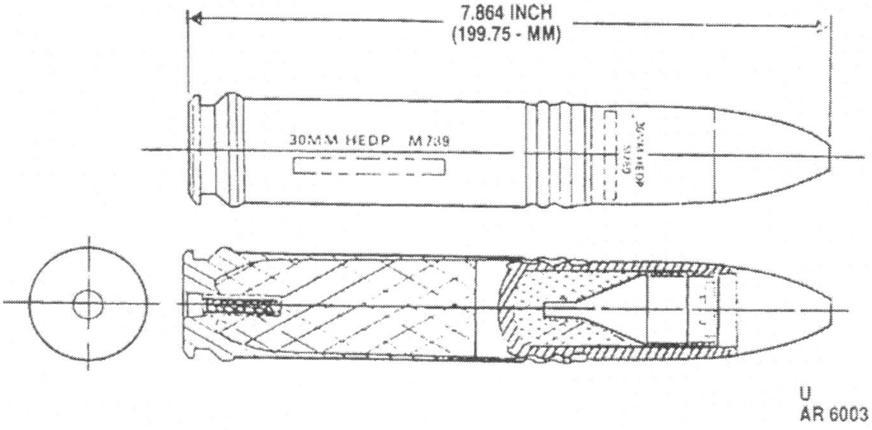


**Type Classification:**

STD.

**Use:**

Automatic cannon, 30mm, M230 (U.S.).

**Description:**

HIGH EXPLOSIVE DUAL PURPOSE Cartridge. The cartridge consists of a steel projectile body loaded with HMX explosive and spin compensated shaped charge liner, point detonating (PD) bore safe fuze and aluminum cartridge case.

**Purpose:**

The projectile fuze arms in flight. Upon impact, the fuze initiates the projectile explosive filler. Detonation of the filler charge collapses the shaped charge liner resulting in the formation of an armor-piercing jet. In addition, main charge detonation produces fragmentation of the projectile body resulting in antipersonnel effects in target vicinity.

**Tabulated Data:**

DODAC ........................................ 1305-B129
UNO serial number ........................ 0321
UNO proper shipping
name ........................................ Cartridges for weapons, with bursting charge
Weight ........................................ 5371 gr
Length ........................................ 7.864 in. (199.75 mm)
Tracer ........................................ NA
Primer........................................ Electric, PA520
Fuze ........................................ PD, M759
Explosive:
Type ........................................ PBXN-5
Weight ........................................ 340 gr
Incendiary:
Type ........................................ NA
Weight ........................................ NA
Propellant:
Type ........................................ WC 855
Weight ........................................ MBR, 50 gr-

**Performance:**

Chamber pressure (avg).................. 40,000 to 44,950 psi (280-310 MPa)
Velocity ........................................ 2640 fps (805 mps)

**Shipping and Storage Data:**

Quantity-distance class/
SCG ........................................ 1.2E (0.4)
Storage code .................................. Class V
DOT shipping class ........................ A

DOT designation ............................ AMMUNITION FOR CANNON WITH SOLID PROJECTILES
Drawing number ............................ 28114062

**References:**

SB 700-20
TM 9-1300-206

## CARTRIDGE, 30MM, DUMMY, M848

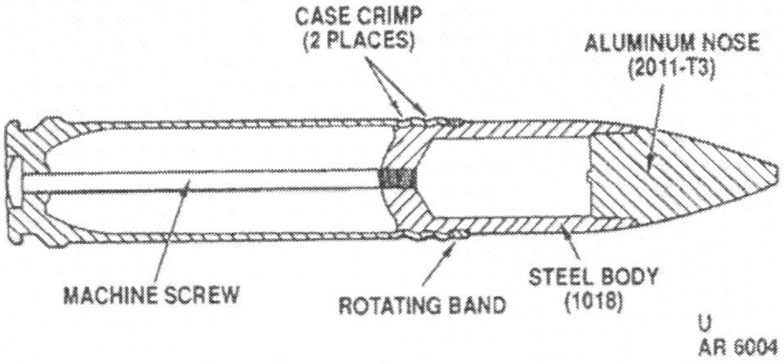


**Type Classification:**

STD - LCC-A MSR 04826009.

**Use:**

Automatic cannon, 30mm, M230. The cartridge is used in drills of the weapon mechanism and to test link/delink functions.

**Description:**

DUMMY Cartridge. The cartridge is completely inert with modified target practice (TP) projectile and cartridge case. Primer and propellant are replaced with threaded steel bolt attached to the base of the projectile body to maintain same weight and balance as the TP cartridge. The projectile body is steel with an aluminum nose.

**Tabulated Data:**

DODAC .......... 1305-B134
Weight .......... 5445 gr
Length .......... 199.75 mm
Tracer .......... NA
Primer .......... NA
Fuze .......... NA

Explosive:
Type .......... NA
Weight .......... NA
Incendiary:
Type .......... NA
Weight .......... NA
Propellant:
Type .......... NA
Weight .......... NA
Projectile:
Length .......... 104.28 mm
Case .......... Aluminum

**Performance:**

Chamber pressure .......... NA
Velocity .......... NA

**Shipping and Storage Data:**

Quantity-distance class/
SCG .......... NA
Storage code .......... NA
DOT shipping class .......... NA
DOT designation .......... NONEXPLOSIVE AMMUNITION
Drawing number .......... 11825935

**References:**

DARCOM 700-3-2

THIS PAGE INTENTIONALLY LEFT BLANK

# CARTRIDGE, 30MM, TEST, M883

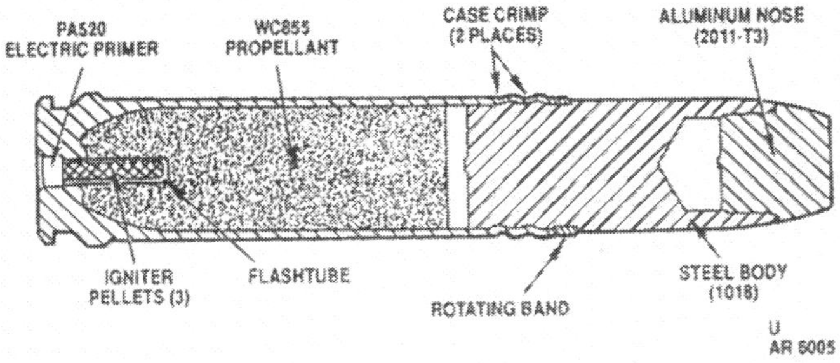


## Type Classification:

STD - LCC-A, MSR 06827003.

## Use:

Automatic cannon, 30mm, M230. The cartridge is used in proof testing the weapon and barrel.

## Description:

TEST Cartridge. The cartridge is a high pressure test round which provides 120 percent chamber pressure to proof test the weapon and barrel. Case, primer, and propellant are identical to the target practice round. Excess pressure is achieved through a heavier 1018 steel projectile. Ogive is blunt and cartridge case is silvered to distinguish from service rounds.

## Tabulated Data:

| | |
|---|---|
| DODAC | 1305-B135 |
| UNO serial number | 0321 |
| UNO proper shipping name | Cartridges for weapons, with bursting charge |
| Weight | 6821 gr |
| Length | 174.35 mm |
| Tracer | NA |
| Primer | Electric, PA520 |
| Fuze | NA |
| Explosive: | |
| Type | NA |
| Weight | NA |
| Incendiary: | |
| Type | NA |
| Weight | NA |
| Propellant: | |
| Type | WC 855 |
| Weight | 772 gr |
| Projectile: | |
| Weight | 5031 gr |
| Length | 79.13 mm |

## Performance:

| | |
|---|---|
| Chamber pressure | NA |
| Velocity | NA |

## Shipping and Storage Data:

| | |
|---|---|
| Quantity-distance class/ SCG | 1.4C |
| Storage code | Class V |
| DOT shipping class | B |
| DOT designation | AMMUNITION FOR CANNON WITH SOLID PROJECTILES |
| Drawing number | 11895941 |

## References:

DARCOM 700-3-2

THIS PAGE INTENTIONALLY LEFT BLANK

# CHAPTER 17

# POINT-DETONATING FUZES

THIS PAGE INTENTIONALLY LEFT BLANK

## FUZE, POINT-DETONATING: M505A3

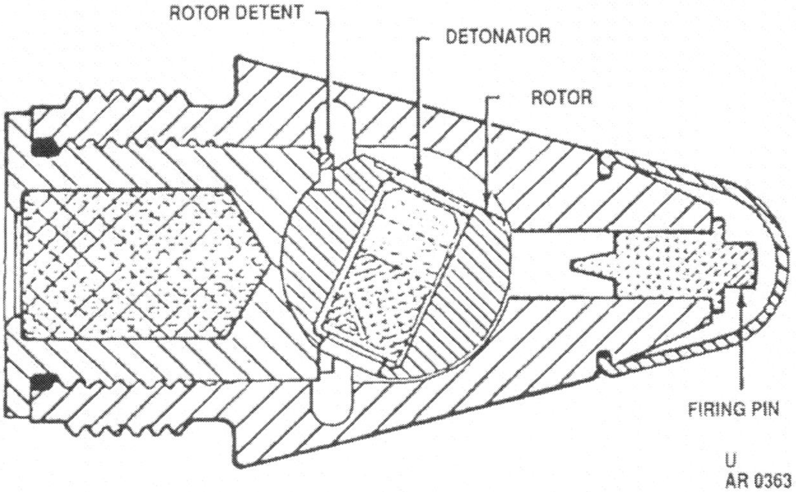


### Type .Classification:

STD.

### Use:

The fuze is designed for use with the 20mm high explosive incendiary (HEI) and high explosive incendiary with tracer (HEI-T) cartridges against aircraft and light armor.:

### Description:

The point-detonating, single-action nose fuze is intended to function upon impact with the target. The fuze is used with the following HEI cartridges: M56A3/A4, M97A2, M210, M242 and M246.

### Function:

Upon impact, the fuze initiates the HE filler and disperses the incendiary filler:. The fuze arms in flight. Its delay is 10 to 35 feet.

### Tabulated Data:

Length (max) ............................ 1.23 in. (31 mm)
Weight ..................................... 330 gr
Diameter .................................. 0.669 in. (17 mm)
Explosive components:
  Detonator ............................... M47
  Booster ................................. T42

### Packing:

For shipping fuzes from the fuze manufacturer to the cartridge LAP facility, the fuzes are packed 200 fuzes per carton, six cartons per wooden box, 1200 fuzes pert wooden box.

### Shipping and Storage:

Quantity-distance class ............ 1.1
DOT shipping class .................. C
DOT designation ........................ DETONATING FUZES, CLASS C EXPLOSIVES
Drawing number ....................... 7258863

THIS PAGE INTENTIONALLY LEFT BLANK

## FUZE, POINT-DETONATING, SELF-DESTROYING: M758

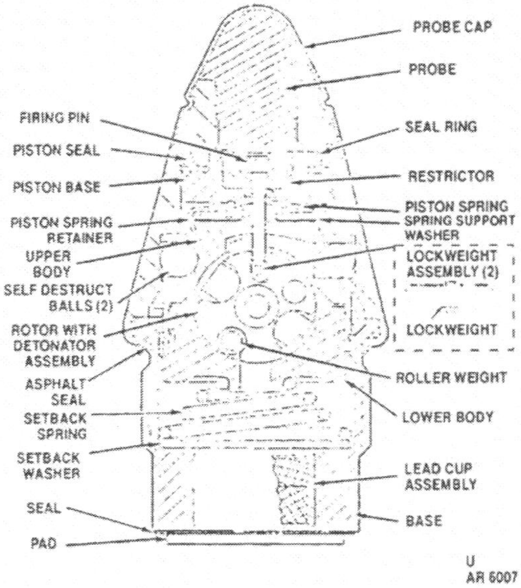


### Type Classification:

STD - MSR 12796010.

### Use:

The fuze is designed for use with the 25mm, high explosive incendiary with tracer, M792 cartridge, fired from the M242 cannon.

### Description:

The point-detonating (PD), self-destroying (SD) M758 fuze is a delayed arming, PD SD type which also features graze impact sensitivity against hard and soft ground media as well as super-quick functioning upon target impact. The fuze consists of an ogive assembly crimped to a base assembly which houses a piston assembly, a piston spring, a body assembly, and a setback spring. The ogive assembly consists of a nylon probe within a steel ogive crimped to a steel probe cap while the base assembly consists of a lead charge within a steel base. the piston spring separates the piston assembly, consisting of an aluminum piston base, a piston seal, a seal ring, a steel restrictor, and a steel firing pill from the body assembly consisting of a nylon upper body, an aluminum lower body, a steel roller weight, a body, a zinc rotor; a detonator, two steel lockweights and lockweight springs which, in turn, separate from the base assembly by the setback spring.

### Function:

Starting from the safe condition, acceleration forces, associated with projectile launch, cause the body assembly and piston assembly located within the fuze ogive to be simultaneously set back to the rear of the fuze against the setback spring. Air originally in the bottom of the fuze is forced through ports into the chamber above the piston. The high spin forces, which occur almost simultaneously with the setback forces, serve three purposes. The spin effect causes two lockweights to move against their firing pin which keeps the motor in a safe position. Spin also causes two balls to move outward into groove located around the inner part of the ogive, locking the body assembly rearward against the compressed force of the setback spring. Spin also causes a rubber piston seal to flare out against the bore of the ogive, resulting in necessary piston scaling action. As the projectile leaves the muzzle, the acceleration force dissipates and the piston becomes free to move forward under forces exerted by the piston spring. The rate of piston movement forward is controlled by air being forced through a porous metal restrictor. This action provides the required arming delay time to meet the fuze no-arm and all-arm requirements. As the piston approaches the forward position, the firing pin is withdrawn from the rotor, allowing it to rotate to the armed position. Centrifugal force acting on the roller weight causes it to move into a groove and lock the rotor in the armed position. The fuze is fully armed when the detonator is in line with the fitting pin and lead assembly. The fuze functions under conditions of direct impact, graze impact or self destruct (non-impact). Direct impact with the target causes a super-quick function of the fuze due to the crushing action of the nose probe. The nose probe is driven into the piston assembly, moving it rapidly rearward. The firing pin, as part of the piston assembly, is thereby driven into the detonator and the explosive train is initiated. Function at low graze angles can occur in two ways. On soft targets where the projectile tends to enter the target, the nose probe is crushed driving the piston firing pin into the detonator. On low angle or graze impacts on hard targets, deceleration of the projectile and spin decay occurs.

Spin decay reduces the centrifugal forces acting on the self-destruct balls and inertial force acts on the body assembly due to projectile deceleration. This, coupled with the stored energy of the setback spring, overrides the force exerted by the self destruct balls and the body assembly is driven forward rapidly. The detonator contacts the filing pill and begins initiation of the explosive train. If no contact is made with a target, the fuze will cause the projectile to function by self-destruct action after a predetermined flight time. Projectile spill decay reduces the centrifugal force acting on the self-destruct balls. The stored energy in the setback spring overrides the holdback force of the balls and forces the body assembly forward to drive the detonator into the firing pin. Initiation of the explosive train occurs, and the projectile is detonated while still in flight.

**Tabulated Data:**

Fuze M758:
  DODAC.................................. IBA
  Length (max) ........................1.417 in. (36 n1ni)
  Weight (approx).....................0.041 lb (18.56 g)
  Diameter (max)......................0.661 in. (16.8 m1m)
  Center of gravity (from
    base) ................................0.602 in. (15.3 3m1m)
Explosive components:
  Detonator ..............................RDX/Lead Axide/
                                                NOL 130, 0.038 g
  Lead charge ..........................HMX, 0.285 g

**Packing:**

For shipping fuzes from the fuze manufacturer to the cartridge LAP facility, one packing unit consists of 1000 fuzes individually inserted into foam trays (200 fuzes per tray and five trays per packing unit) which are stacked with chipboard separators and then packed into inner and outer fiberboard boxes.

Length .....................................19 in. (482.6 1mm)
Width ......................................12.5 in. (317.5 mm)
Height .....................................10.1 in. (25G6.5 11m)
Weight (empty).........................11.5 I) (5.23 kg)
Weight (1000 fuzes)..................52.6 lb (23.86 kg)

**Shipping and Storage Data:**

Quantity-distance class/
  SCG .....................................1.1B
Storage code ............................Class VI
DOT shipping class...................A
DOT designation ......................DETONATING
                                                FUZES, CLASS A
                                                EXPLOSIVES
DOCAC....................................TBA
Drawing number .......................DS8861

**References:**

DARCOM 700-3-2
TM 9-1300-206

## FUZE, POINT-DETONATING: M759 (XM714E6)

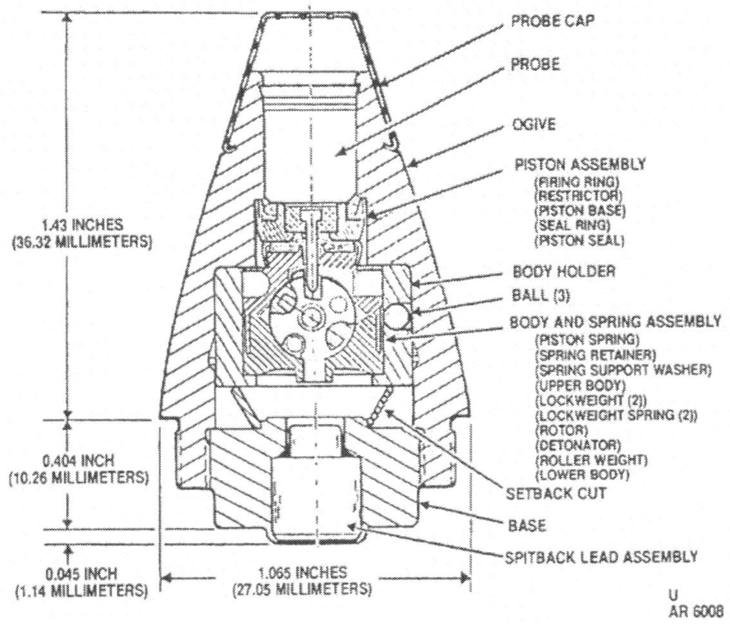


**Type Classification:**

STD

**Use:**

The fuze is designed for use with the 30mm high explosive dual purpose (HEDP) cartridges fired from the M230 automatic cannon.

**Description:**

Fuze M759 (formally designated as the XM714E6) is a point-detonating type with detonator and spitback charge. It is used with HEDP cartridge M789.

**Function:**

Prior to launch, the fuze detonator is housed in the rotor with the rotor locked in the safe position by the firing pin tip and two lockweights. On setback, the piston assembly (with firing pin) and body assembly (with detonator and rotor) are moved to the most rearward position by the firing pill tip and two lockweights. As the fuzed projectile exits the gun muzzle, spin environment causes the lockweights to retract from the rotor and the two steel balls to engage the ogive groove, locking the body assembly in the position it had achieved at setback. In addition, spin causes the piston seal to flare and engage the ogive sidewalls which eliminates the air flow path between the ogive and piston. Movement of the piston assembly (firing pin) to the forward or armed position is delayed due to the regulated flow of trapped air through the restrictor until the projectile has traveled a minimum of 15 meters from the muzzle. After retraction to the firing pin from the rotor (last safety lock), the spin causes the rotor (with detonator) to rotate to the armed state, i.e., stab detonator aligned with firing pin and lead cup assembly. As the rotor achieves the armed position (150 in), the roller weight moves outboard from the rotor and engages the detent from the lower and upper bodies thus locking the rotor in armed state. Upon target impact, the probe nose piece is driven into the ogive forcing the piston assembly (firing pin) aft. The firing pin stabs the detonator and its explosive initiates the lead cup assembly explosive which, in turn, ignites the main charge.

**Tabulated Data:**

| | |
|---|---|
| Type | PD |
| Weight | 1.41 oz (40 g) |
| Length | 1.88 in. (47.72 mm) |
| Diameter | 1.06 in. (27.05 mm) |
| Material | Aluminum |
| Non-arm range | 15 ii |
| All-arm range | 150 m |
| Explosive components: | |
| Detonator | NOL No. 130 Priming Mix, 0.0055 g |
| Lead Azide | 0.015 g |
| RDX Type B | 0.016 g |
| Spitback lead assembly: | |
| PBXN, Type II, C1, 2 or 3 | 0.25 g |
| PBXN-5, Type I, C1 or 3 | 0.06 g |

**Shipping and Storage Data:**

| | |
|---|---|
| Quantity-distance class/ SCG | 1.IB |
| DOT shipping class | A |
| DOT designation | DETONATING FUZES, CLASS A EXPLOSIVE |
| Drawing number | 11834312 |

# CHAPTER 18

# ACCESSORIES

## GENERAL

a. Accessories. Accessories used with small arms ammunition such as clips and metallic belt links, are described in this chapter. These items are listed and packed as indicated in SC 1305/30-IL and also listed in TM 9-1305-20120&P

b. Magazines for Pistols and Other Small Arms. Magazines for pistols and other small arms do not come preloaded from the ammunition supply point. The magazine for the submachine gun M3 and M3A1 holds 30 rounds of ammunition and this also is not preloaded.

c. Metallic Link Belts. Ammunition for use in machine guns is issued in metallic link belts. Link belts are made of units of cartridges and links. Each link has two loops fitted around a single cartridge and one loop fitted around an adjacent cartridge. Each cartridge in a metallic link belt has two links attached to it, except the end cartridges. Cartridge links are made of steel which is processed to prevent rusting. They are manufactured and tested to assure satisfactory ammunition feeding and functioning under all service conditions.

THIS PAGE INTENTIONALLY LEFT BLANK

## ACCESSORIES FOR CARTRIDGES, CALIBER .50

**Links, Closed Loop, M2 and M9:**

Caliber .50, M2 and M9 links (fig. 18-7) are manufactured with closed loops. When assembled with the cartridge in belts, the links fit on the cartridge shoulder. The tapered front loops of the links are positioned firmly on the cartridge shoulder to hold the cartridge in proper alignment for feeding into the weapon. The design required the cartridge to be extracted from the rear and dropped into position for moving into the chamber for firing. Weapons designed for the links require additional space in the rear of the receiver for retraction. M2 and M9 links are used with the M2 machine gun only.
Link, M2 (NSN 1305-00-096-3181(A640))
Link, M9 (NSN 1305-00-078-8463(A641))

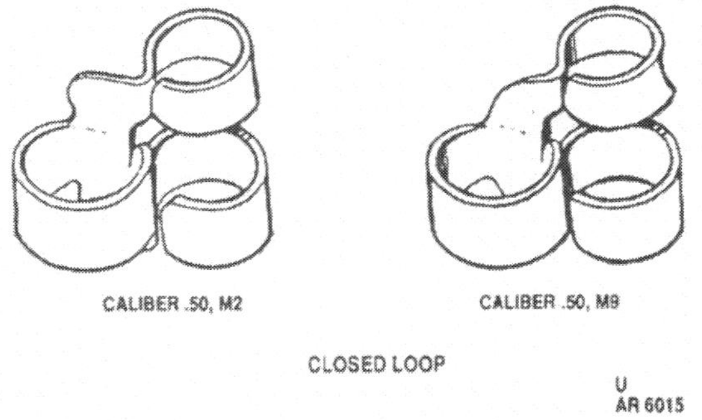


**Figure 18-7. Caliber .50, M2 and M9 links (closed loop).**

**Link, Open Loop, M15A2:**

The caliber .50, M15A2 link (fig. 18-8) is manufactured with partially open loops and has a positioning finger on one side which snaps into the extractor grooves of the cartridge to retain the cartridge in proper feed alignment. The link design permits a portion of bolt to ride through link loop openings and push the cartridges forward and out of the link into the chamber for firing. The M15A2 link is used with the M85 machine gun.

**References:**

SC 1305/30IL
TM 9-1305-201-20&P

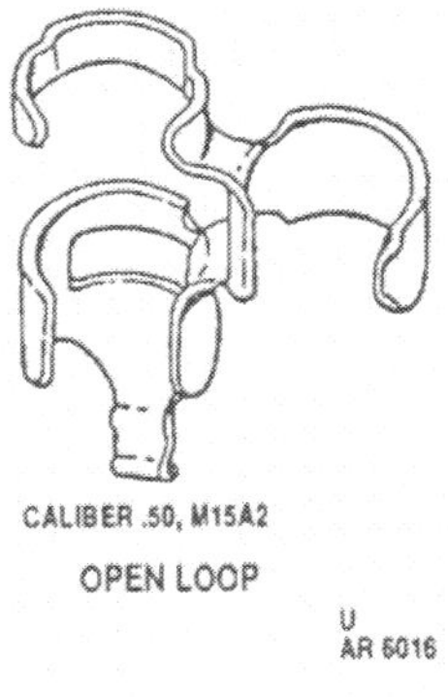


**Figure 18-8. Caliber .50, M15A2 link (open loop).**

THIS PAGE INTENTIONALLY LEFT BLANK

## ACCESSORIES FOR CARTRIDGES, 5.56MM

### Clips:

Cartridges for the 5.56mm M16 rifle are assembled in 10-round clips. The clip (fig. 18-1) consists of a channel-shaped body containing a spring with narrow tabs at each end. After the clip is filled, the tabs are bent at right angles to the extrator rims of the end cartridges to hold them securely in the clip (NSN 1305-00-0524377).

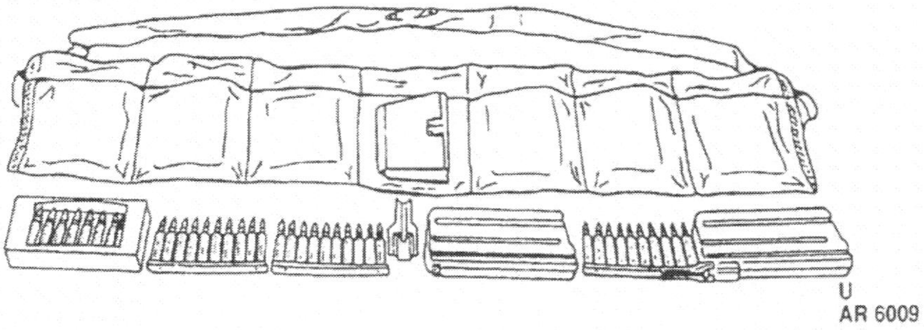


**Figure 18-1. Cartridge clips and bandoleer for 5.56mm cartridges.**

### Link M27:

The M27 link (fig. 18-2) is a reduced size of the 7.62nmm link M13. The body link fits into the extractor groove on the cartridge case. The U.S. Navy, uses the link with the Stoner Machine Gun M63. The U.S. Army designated its use foil the M249 Machine Gun. The link is used with M855 Ball Cartridges, M856 Tracer, and M200 Blank only.

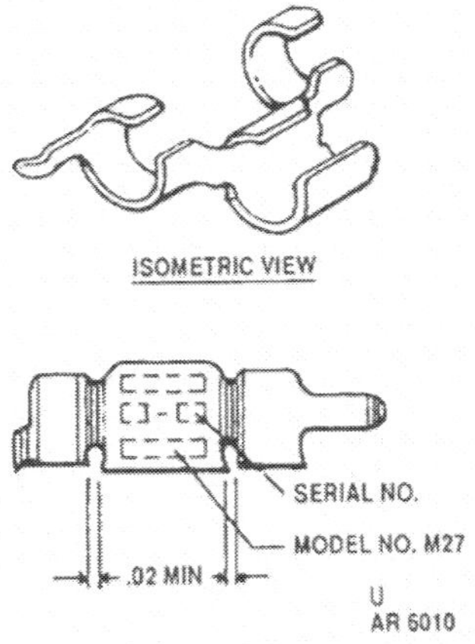


**Figure 18-2. M27 link for 5.56mm.**

### Other Accessories:

Fillet; Magazine.
Bandoleer; M3 (NSN 1305-00-9)24-3088).

### References:

SC 1305/30IL
TM 9-1305-201-20&P

THIS PAGE INTENTIONALLY LEFT BLANK

## ACCESSORIES FOR CARTRIDGES, 7.62MM

### Clips:

Cartridges for the 7.62mm M14 rifles are assembled in 5-round clips (fig. 18-3). All cartridges are packed in bandoleers with separate magazine fillers. The clip, which consists of a channel-shaped body containing a spring, facilitates loading of magazines. Loaded clips are fitted into the top of the rifle receiver and the cartridges are pressed into the magazine assembly. A magazine fillet; which adapts clips to the magazine, allows individual magazines to be loaded with clipped ammunition. The clip is 2.36 inches long. The 7.62mm ammunition and clip are shown with the M2 bandoleer in figure 18-4 and with the M4 bandoleer in figure 18-5. The M4 bandoleer is for belted ammunition for machine guns (NSN 1305-00-914-7912).

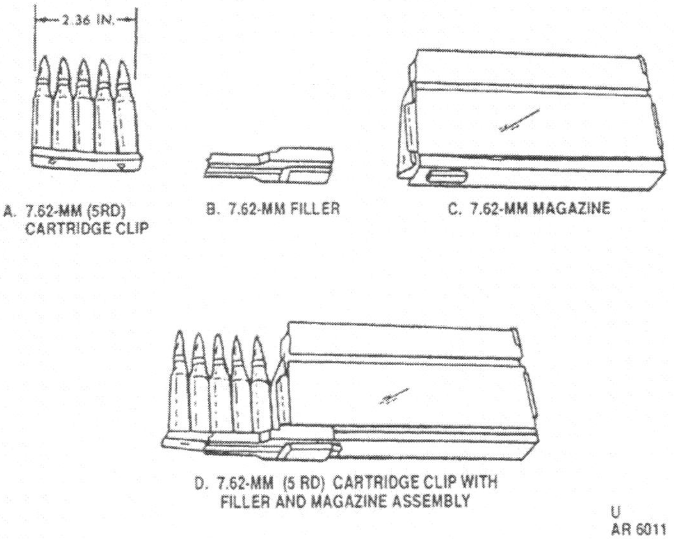


**Figure 18-3. Cartridge clip, filler, and magazine assembly for 7.62mm cartridges.**

**Figure 18-4. M2 bandoleer for 7.621imm cartridges.**

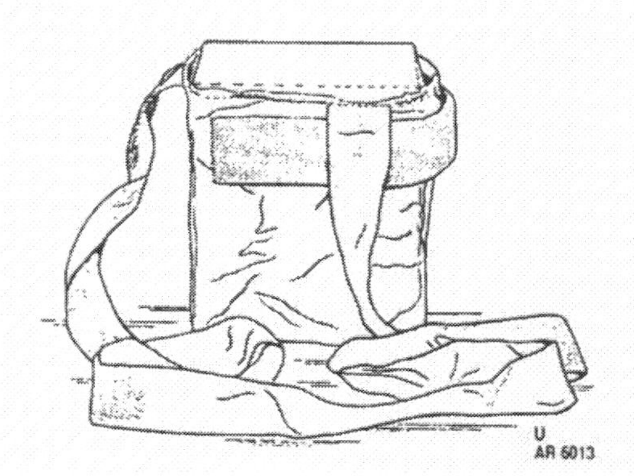


Figure 18-5. M4 bandoleer for 7.62mm cartridges.

**Link M13:**

The 7.62mm M13 link (fig. 18-6) is manufactured with partially open loops and has a positioning finger on one side which snaps into the extractor grooves of cartridges to retain the cartridge in proper feed alignment. The link design permits a portion of the bolt to ride through the link opening and push cartridges forward and out of the link into the chamber for firing.

Weapons using this type link are designed and manufactured with a short receiver, i.e., M60, M73, M219, M240, and M134 machine guns (NSN 1305-00-752-7815).

**Other Accessories:**

Fillet; Magazine.
Bandoleer; M2 (NSN 1305-00-984-1125).
Bandoleer M4 (NSN 1305-00-079-3579).

**References:**

SC 1305/30IL
TM 9-1305-201-20&P

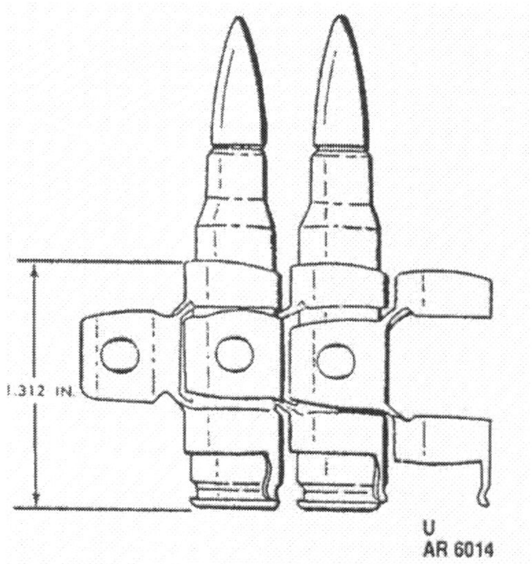


Figure 18-6. Cartridge M13 link for 7.62mm cartridges.

## ACCESSORIES FOR CARTRIDGES, 20MM

**Link M10:**

The M10 link (fig. 18-9) is utilized in the M3 and M24 belt-fed automatic guns. The link is of steel and the belt is formed by the placement of cartridges within the links. The link is self-separating when the cartridges are removed during the gun feeding cycle. The metallic link can be used to make up belts for either right-hand or left-hand feeding of the gun. A cartridge metallic-link filler is used at the end of the belt when it is desired to load belts for either double-loop left-hand feed or double-loop right-hand feed. The purpose of the link-filler is to close the single loop when it is at the end of the belt to prevent snagging in the feed chute (NSN 1305-00-028-4869).

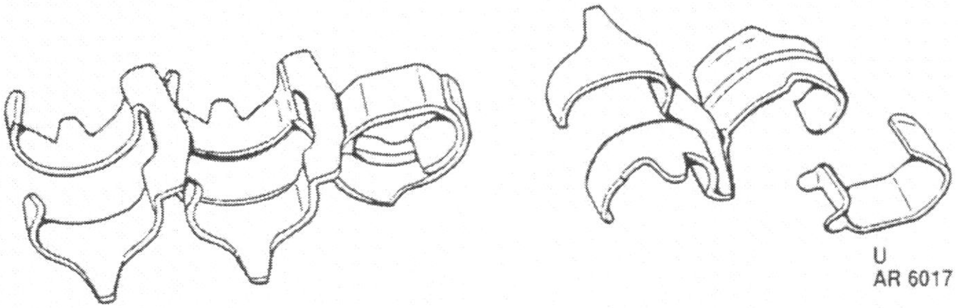


**Figure 18-9. M10 link for 20mm cartridges.**

**Link M12:**

The M12 metallic link (fig. 18-10) is used in belt-fed gas-operated M39 series automatic aircraft guns. The ammunition belt is formed by connecting a specified number of links together and then inserting a cartridge for each link. The link is self-separating when cartridges are removed during the gun-feeding cycle. The M12 is an end loading link and is linked for either the right-hand or0 left-hand feeder (NSN 1305-00-541-0021).

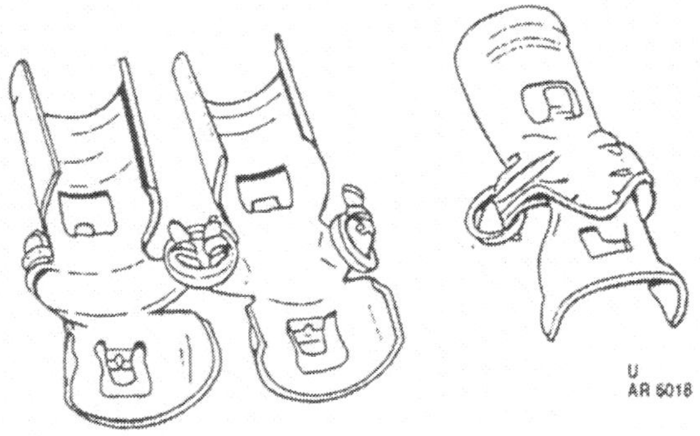


**Figure 18-10. M12 link for 20mm cartridges.**

**Link M14A2:**

The M14A2 link (fig. 18-11) is used primarily for linking 20amm ammunition for belt-fed M61 automatic guns and loading of ammunition into linkless feed systems used in aircraft and ground defense vehicles. The link is self-separating when cartridges are removed during the feeding cycle (NSN 1305-00-935-9219).

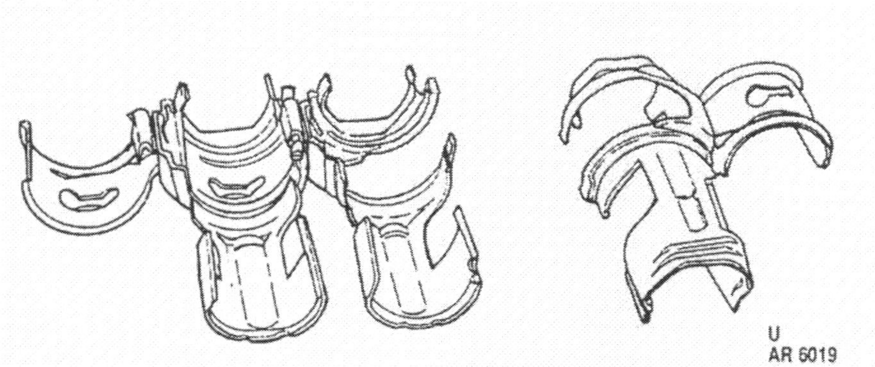


**Figure 18-11. M14A2 link for 20mm cartridges.**

**Link M17A2:**

The M17A2 link (fig. 18-12) is used for linking 20mim ammunition for belt-fed M61 automatic weapon system in the B52 G & H aircraft. The link is self-separating when the cartridge is removed from the link during the feeding cycle,

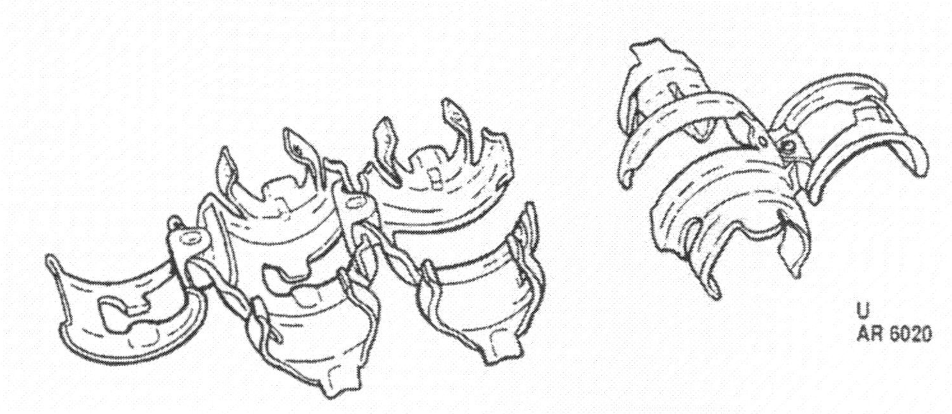


**Figure 18-12. M17A2 link for 20mm cartridges.**

**Link M24 and M22:**

The M24 link (fig. 18-13) is a metallic link for use with the following 20mm cartridges: M206A1 Target-Practice, M599 High-Explosive Incendiary Tracer, and M601 Armor-Piercing Incendiary racer. The European (German) DMI/DMIA link is also used for these cartridges. The DMI/DMIA link is used with the HSH20 seriers. The M22 link is for U.S. Air Force use only. It is used to upload their linkless system. It is reusable and does not disintegrate.

Link, M22 (NSN 1305-00-926-1939).

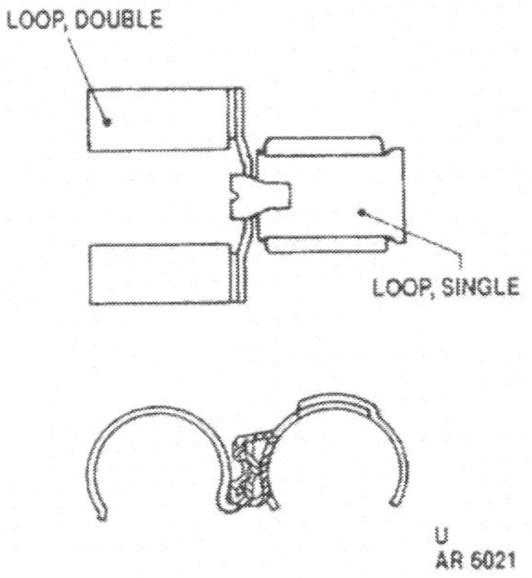


**Figure 18-13. M24 link.**

THIS PAGE INTENTIONALLY LEFT BLANK

## ACCESSORIES FOR CARTRIDGES, 25MM

**Link M28:**

The M28 link (fig. 18-14) is a two piece assembly intended to hold and link the 25mim ammunition for guiding the ammunition through the weapon system. The assembly consists of hardened steel main and pivot links hooked together and contoured accordingly for its intended purpose. The Swiss Oerlikon link and the M28 male dimensionally identical.

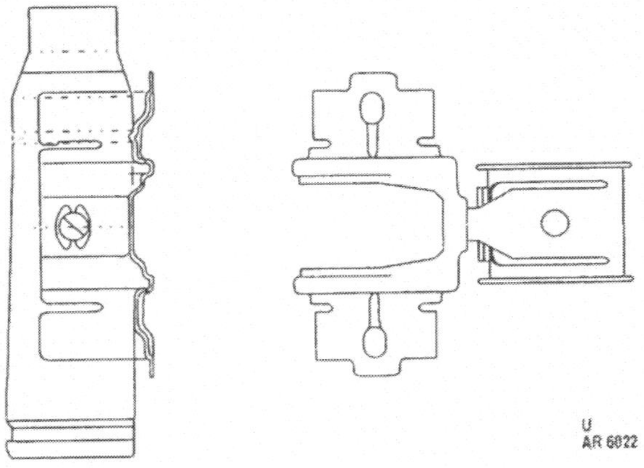


**Figure 18-14. M28 link for 25mm cartridges.**

THIS PAGE INTENTIONALLY LEFT BLANK

ACCESSORIES FOR CARTRIDGES, 30MM

**Link M29:**

The M29 link (fig. 18-15) is 1050 steel with type Z (link) phosphate coating. It is used for interoperability with the NATO ADEN/DEFA weapon systems) (NSN 1305-01-078-5505).

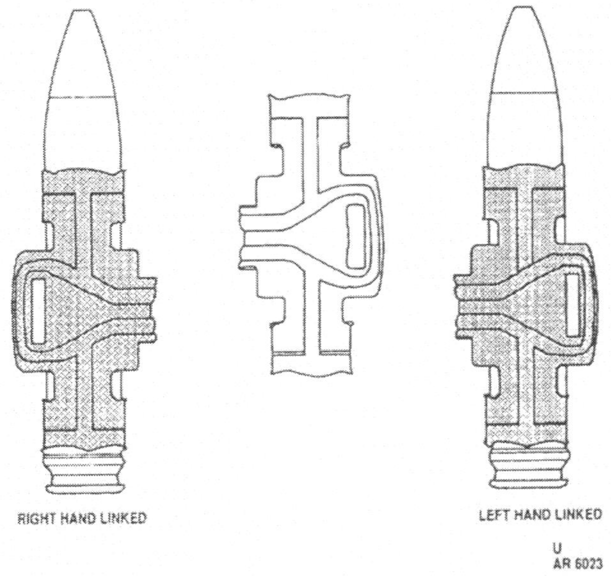


**Figure 18-15. M29 link.**

THIS PAGE INTENTIONALLY LEFT BLANK

## TABLE 18-1. SMALL ARMS AMMUNITION

| Caliber | Ammunition | Clip | Magazine | Link | Weapon used |
|---|---|---|---|---|---|
| 5.56mm | M193 Ball<br>M195 Grenade<br>M196 Tracer<br>M197 High Pressure Test<br>M199 Dummy<br>M200 Blank | 10 rd | 20 and 30 rd | | M231 Submachine Gun for M196 tracer only. M16/ M16A1 rifle. |
| | M855 Ball<br>M856 Tracer | | | M27 | Stoner M63 (Navy) and M249 machine guns. |
| 7.62mm | M59, M80 Ball<br>M61 Armor Piercing<br>M62 Tracer<br>M63 Dummy<br>M82 Blank<br>M60 High Pressure Test | 5 rd | 20 rd | M13 open loop for MG | M14 rifle; M60, M73, M219, M240*, and M134 machine guns. |
| | M64 Grenade | | | | M14 rifle only. |
| | M118 Ball Match | | | | M14 rifle only |
| | M160 Frangible | | | | M219 MG only. |
| | M172 Dummy | | | | M60 MG only. |
| | M852 Ball Match | | | | |

NOTE
10 rd clip f/M16A1.
5 rd clip f/M14.
20 rd mag f/M16A1.
30 rd mag f/M16Al.
20 rd mag f/M14.

| Caliber | Ammunition | Clip | Magazine | Link | Weapon used |
|---|---|---|---|---|---|
| Cal. 45 | Ball, M1911 Blank<br>M9 High Pressure Test M1 | | 7 rd for pistol<br>30 rd for sub-machine gun. | N/A | Pistol M1911 andM19llAl. Submachine gun M3 and M3A1. |
| Cal .50 | M2 Ball<br>M33 Ball<br>M2 Armor Piercing<br>M8 Armor Piercing Incendiary | N/A | N/A | **M15A2 open loop with M85 MG. | M85 Machine Gun. |
| | M20 Armor Piercing Incendiary-Tracer<br>M1 Incendiary<br>M23 Incendiary<br>M1 Tracer<br>M10 Tracer<br>M17 Tracer<br>M2 Dummy, High Pressure Test | N/A | N/A | ***M2 and M9 closed loops with M2 MG. | M2 Machine Gun only. |
| | M1 Blank<br>M1A1 Blank<br>with M2 MG. | N/A | N/A | ***M2 and M9 closed loops | M2 Machine Gun only. |
| | M48A2 Spotter-Tracer | N/A | 10 and 20 rd | N/A | M8C Spotting Rifle |
| | T249E2 Practice | N/A | 10 and 20 rd | N/A | M8C Spotting Rifle |
| | T251 High Pressure Test | N/A | 10 and 20 rd | N/A | M8C Spotting Rifle |

* M240 Machine Gun only fires the Ball M80 and the Tracer M62.
** Link M15A2 Open Loop used with M85 Machine Gun only.
***Link M2 and M9 Closed Loop used with M2 Machine Gun only.

**TABLE 18-1. SMALL ARMS AMMUNITION - Continued**

| Caliber | Ammunition | Link | Weapon (gun) |
|---|---|---|---|
| 20mm | HEI-T-SD M246/M246A1<br>HEI M56A3/M56A4<br>TP-T M220<br>TP M55A2<br>Dummy M51A2 | M14 series | M167A1 (M168 Cannon) |
| 20mm | API-T M52E1<br>API M53<br>HEI M56A3/M56A4<br>HEI-T M242<br>HEI-T-SD M246/M246A1<br>TP M55A2<br>TP-T M220<br>Dummy M51A2 | M12 Metallic Link for M39 Cannon.<br>M14A2 for M61 Cannon. | M39, M61 |
| 20mm | API-T M52E1<br>API M53<br>HEI M56A3/M56A4<br>HEI-T M242/M242A1<br>HEI-T-SD M246/M246A1<br>TP M55A2<br>TP-T M220<br>Dummy M51A2 | M17A2 | M61 |
| 20mm | API M53<br>HEI M56A3/M56A4<br>TP M55A2<br>Dummy M51A2 | M14A2 | M195<br>Linker-Delinker M25 |
| 20mm | HEI-T-SD M246/M246A1<br>HEI M56A3/M56A4<br>TP-T M220<br>TP M55A2<br>Dummy M51A2 | M14A2 | 168, 197 |
| 20mm | AP-T M95<br>Incendiary M96<br>TP M204<br>HEI M210 | M10 | M3 |
| 20mm | AP-T M95<br>Incendiary M96<br>HEI M97A2<br>TP M99A1 | M10 | M24A1 |
| 20mm | HEI-T M599<br>API-T M601<br>TP-T M220 | M24 or DMI/DMIA | 139 |
| 25mm | API-T M791<br>HEI-T M792<br>TP-T M793 | M28 | M242 |
| 30mm | HEDP M789<br>TP M788<br>HEI M799 | M29 | M230<br>NATO M552/M553 |

*U.S. GOVERNMENT PRINTING OFFICE: 1995 - 646-049/00715

Order of the Secretary of the Army:

GORDON R. SULLIVAN
General, United States Army
Chief of Staff

Official:

MILTON H. HAMILTON
Administrative Assistant to the
Secretary of the Army
06732

Distribution:
To be distributed in accordance with DA Form 12-34E, Block 0847, requirements for IM 43-0001-27.

*U.S. G.P.O.:1994-546-043:80742

RECOMMENDED CHANGES TO EQUIPMENT TECHNICAL PUBLICATIONS

**SOMETHING WRONG WITH PUBLICATION**

FROM: (PRINT YOUR UNIT'S COMPLETE ADDRESS)

DATE SENT

| PUBLICATION NUMBER | PUBLICATION DATE | PUBLICATION TITLE |
| --- | --- | --- |
| | | |

| BE EXACT PIN-POINT WHERE IT IS | | | | IN THIS SPACE, TELL WHAT IS WRONG AND WHAT SHOULD BE DONE ABOUT IT. |
| --- | --- | --- | --- | --- |
| PAGE NO. | PARA-GRAPH | FIGURE NO. | TABLE NO. | |
| | | | | |

| PRINTED NAME, GRADE OR TITLE AND TELEPHONE NUMBER | SIGN HERE |
| --- | --- |
| | |

DA FORM 1 JUL 79 2028-2

PREVIOUS EDITIONS ARE OBSOLETE.

P.S.--IF YOUR OUTFIT WANTS TO KNOW ABOUT YOUR RECOMMENDATION MAKE A CARBON COPY OF THIS AND GIVE IT TO YOUR HEADQUARTERS.

## The Metric System and Equivalents

### *Linear Measure*

1 centimeter = 10 millimeters = .39 inch
1 decimeter = 10 centimeters = 3.94 inches
1 meter = 10 decimeters = 39.37 inches
1 dekameter = 10 meters = 32.8 feet
1 hectometer = 10 dekameters = 328.08 feet
1 kilometer = 10 hectometers = 3,280.8 feet

### *Weights*

1 centigram = 10 milligrams = .15 grain
1 decigram = 10 centigrams = 1.54 grains
1 gram = 10 decigram = .035 ounce
1 decagram = 10 grams = .35 ounce
1 hectogram = 10 decagrams = 3.52 ounces
1 kilogram = 10 hectograms = 2.2 pounds
1 quintal = 100 kilograms = 220.46 pounds
1 metric ton = 10 quintals = 1.1 short tons

### *Liquid Measure*

1 centiliter = 10 milliters = .34 fl. ounce
1 deciliter = 10 centiliters = 3.38 fl. ounces
1 liter = 10 deciliters = 33.81 fl. ounces
1 dekaliter = 10 liters = 2.64 gallons
1 hectoliter = 10 dekaliters = 26.42 gallons
1 kiloliter = 10 hectoliters = 264.18 gallons

### *Square Measure*

1 sq. centimeter = 100 sq. millimeters = .155 sq. inch
1 sq. decimeter = 100 sq. centimeters = 15.5 sq. inches
1 sq. meter (centare) = 100 sq. decimeters = 10.76 sq. feet
1 sq. dekameter (are) = 100 sq. meters = 1,076.4 sq. feet
1 sq. hectometer (hectare) = 100 sq. dekameters = 2.47 acres
1 sq. kilometer = 100 sq. hectometers = .386 sq. mile

### *Cubic Measure*

1 cu. centimeter = 1000 cu. millimeters = .06 cu. inch
1 cu. decimeter = 1000 cu. centimeters = 61.02 cu. inches
1 cu. meter = 1000 cu. decimeters = 35.31 cu. feet

## Approximate Conversion Factors

| *To change* | *To* | *Multiply by* | *To change* | *To* | *Multiply by* |
|---|---|---|---|---|---|
| inches | centimeters | 2.540 | ounce-inches | Newton-meters | .007062 |
| feet | meters | .305 | centimeters | inches | .394 |
| yards | meters | .914 | meters | feet | 3.280 |
| miles | kilometers | 1.609 | meters | yards | 1.094 |
| square inches | square centimeters | 6.451 | kilometers | miles | .621 |
| square feet | square meters | .093 | square centimeters | square inches | .155 |
| square yards | square meters | .836 | square meters | square feet | 10.764 |
| square miles | square kilometers | 2.590 | square meters | square yards | 1.196 |
| acres | square hectometers | .405 | square kilometers | square miles | .386 |
| cubic feet | cubic meters | .028 | square hectometers | acres | 2.471 |
| cubic yards | cubic meters | .765 | cubic meters | cubic feet | 35.315 |
| fluid ounces | milliliters | 29,573 | cubic meters | cubic yards | 1.308 |
| pints | liters | .473 | milliliters | fluid ounces | .034 |
| quarts | liters | .946 | liters | pints | 2.113 |
| gallons | liters | 3.785 | liters | quarts | 1.057 |
| ounces | grams | 28.349 | liters | gallons | .264 |
| pounds | kilograms | .454 | grams | ounces | .035 |
| short tons | metric tons | .907 | kilograms | pounds | 2.205 |
| pound-feet | Newton-meters | 1.356 | metric tons | short tons | 1.102 |
| pound-inches | Newton-meters | .11296 | | | |

## Temperature (Exact)

| | | | | |
|---|---|---|---|---|
| °F | Fahrenheit temperature | 5/9 (after subtracting 32) | Celsius temperature | °C |

PIN: 049093-000

Made in United States
Troutdale, OR
11/03/2026

56847584R00182